Nanotechnology and Scientific Communication

Deborah R. Bassett

Nanotechnology and Scientific Communication

Ways of Talking about Emerging Technologies and Their Impact on Society (2004–2008)

Deborah R. Bassett
Department of Communication
University of West Florida
Pensacola, Florida, USA

ISBN 978-1-349-95200-7 ISBN 978-1-349-95201-4 (eBook)
DOI 10.1057/978-1-349-95201-4

Library of Congress Control Number: 2017937317

© The Editor(s) (if applicable) and The Author(s) 2017
This work is subject to copyright. All rights are solely and exclusively licensed by the Publisher, whether the whole or part of the material is concerned, specifically the rights of translation, reprinting, reuse of illustrations, recitation, broadcasting, reproduction on microfilms or in any other physical way, and transmission or information storage and retrieval, electronic adaptation, computer software, or by similar or dissimilar methodology now known or hereafter developed.
The use of general descriptive names, registered names, trademarks, service marks, etc. in this publication does not imply, even in the absence of a specific statement, that such names are exempt from the relevant protective laws and regulations and therefore free for general use.
The publisher, the authors and the editors are safe to assume that the advice and information in this book are believed to be true and accurate at the date of publication. Neither the publisher nor the authors or the editors give a warranty, express or implied, with respect to the material contained herein or for any errors or omissions that may have been made. The publisher remains neutral with regard to jurisdictional claims in published maps and institutional affiliations.

Cover image © Bitboxx.com

Printed on acid-free paper

This Palgrave Macmillan imprint is published by Springer Nature
The registered company is Nature America Inc.
The registered company address is: 1 New York Plaza, New York, NY 10004, U.S.A.

Acknowledgements and Disclaimer

I would like to thank the University of West Florida Communication Department, especially Dr. Kurt Wise, Dr. Athena du Pre, Pamela Harlin, and Shelby Villatoro Calderon, for their support during the completion of this project.

This material is based upon work supported by the National Science Foundation under Grant No. 0335765. Any opinions, findings, and conclusions or recommendations expressed in this material are my own and do not necessarily reflect the views of the National Science Foundation.

Contents

1 Introduction and Overview — 1

2 Discovering a Code of Scientific Communicative Conduct — 23

3 Ways of Speaking about Social Relations — 55

4 Strategic Conduct in a Code of Science — 77

5 Evidence for Multiple Speech Codes — 115

6 Conclusions — 131

Appendix A: Transcription Conventions — 141

Bibliography — 143

Index — 153

CHAPTER 1

Introduction and Overview

Abstract In this chapter, the history of nanotechnology and the theoretical perspectives that guided and informed the research, including cultural communication theory and speech codes theory, are presented and discussed. The goal in conducting this research was to understand the cultural aspects of the discourse of nanoscience/nanotechnology through a comparison of observed behavior and actual speech. To achieve this goal multiple modes of inquiry were used, including fieldwork, participant observation, and interviews, over a period of 4 years.

Keywords Ethnography of communication · Nanotechnology · Social and ethical issues · Speech codes theory

In 1959, Richard Feynman delivered a speech entitled "There's Plenty of Room at the Bottom" to the American Physical Society in which he proposed developing science and technology that would enable manipulation of matter at molecular and atomic levels. Although this speech can be seen as the origin of nanoscale science and engineering (NSE), or nanoscience and nanotechnologies (NS/NT), as a formal proposal, these labels only began to emerge some 25 years later with the publication of K. Eric Drexler's book *Engines of Creation* (1986) in which he introduced the term "nanotechnology" to a mainstream audience.[1]

According to Bennett and Sarewitz (2006), the development of the atomic force microscope[2] and the buckyball by physicists in the 1980s began to realize Feynman's vision and by the early 1990s, the National Science Foundation (NSF) launched its first NSE research program. The 21st Century Nanotechnology R&D Act was passed by US Congress in 2003 and led to the establishment of research centers around the United States. Bennett and Sarewitz trace the development of the nano phenomenon, showing how federal funding for NSE shot up dramatically from $6 million in 1991 to $961 million by 2004 and "nano" began showing up in exponentially increasing numbers in titles of journal articles, journal titles, and grant proposals. For example, the number of articles listed in the Science Citation Index containing the word nano in their titles increased from several hundred in the late 1980s to more than 23,000 in 2003 (Bennett and Sarewitz, 2006) and more than 150,000 in 2008.

SOCIAL AND ETHICAL ISSUES (SEI)

Around the same time that the term nano was appearing with increasing frequency in journals and grant proposals, K. Eric Drexler, an engineer from Massachusetts Institute of Technology (MIT), published *Engines of Creation* (1986), the earliest detailed account of how nanotechnologies might develop. And although Drexler's account was overwhelmingly optimistic, his description of the possible risk of self-assembling nanobots going amuck and attacking the biosphere (a scenario that is referred to as "the gray goo" problem) created concerns about the development of self-assembling molecules, or molecular manufacturing. The "gray goo problem" refers to the following phrase from Drexler's (1986) book:

> Tough, omnivorous "bacteria" could out-compete real bacteria: they could spread like blowing pollen, replicate swiftly, and reduce the biosphere to dust in a matter of days. Among the cognescenti of nanotechnology, this threat has become known as the gray goo problem (Drexler, 1986, p. 172).

These concerns led to consumer and environmental groups such as the Action Group on Erosion, Technology and Concentration (ETC) Group calling for a moratorium on research that would lead to molecular manufacturing (ETC Group, 2003). In 2000 when *Wired Magazine* ran a cover story by then CEO of Sun Microsystems Bill Joy entitled "Why the Future Doesn't Need Us" (Joy, 2000), concern about the impact of nanotechnologies reached a level of noise that began to receive media, public, and scholarly attention. Joy, who was well-known for his contributions to the development of the Internet, Java, and Jini, essentially called for a halt to research and development in genetics, nanotechnology, and robotics that would enable the sort of self-assembling, self-replicating molecules that Drexler (1986) described. Joy quoted Drexler's 1986 reference to gray goo in his essay. The reactions to Joy's essay included more than 10,000 articles in response (Kurzweil, 2005, p. 394) and ranged from accusations that he "focused entirely on the downside scenarios" (Kurzweil, 2005, p. 394) to "one of the great Paul Revere moments of our time, a full-throated and unhesitating alarum that should scare the hell out of us" (McKibben, 2003, p. 92). Joy's article has since become a ubiquitous reference in publications about the impact of nanotechnologies and can generally be credited with beginning the conversation about the potential ramifications of NSE among the social scientific community (Bennett and Sarewitz, 2006).

Although many in the mainstream scientific community rejected Drexler's vision, as well as Joy's concerns, molecular manufacturing was frequently referred to as "the holy grail" of NSE research and development (McKibben, 2003, p. 80; and my own personal communication with NSE scientists and engineers) and the same revolutionary narrative that is applied to NSE by Drexler and others was deployed by the mainstream scientific community (Sparrow, 2007). In the early 2000s, "nanotechnology" became a trendy buzzword among science and technology circles, as well as in popular media, and was introduced into the mass consumer lexicon with the unveiling of the "nano" iPod by Apple in 2005.

DEFINITIONAL ISSUES

Defining nanotechnology is not as easy one might assume. Schummer (2007) offers three different approaches to defining nanotechnologies, with the caveat that "all are conspicuously vague" (p. 293). The first approach Schummer presents is that of a "nominal definition," which "provides necessary and sufficient conditions that a technology must meet to be called a nanotechnology." Schummer notes that "[t]ypical definitions" of nanotechnology refer to a size range of 1–100 nm, but suggests that this definition is problematic as it "covers all the classical natural science and engineering disciplines that investigate and manipulate material objects" (p. 293). The second definitional approach he offers is "teleological definition" which "defines nanotechnology by its future goals." Schummer rejects this definition, saying it is "impracticable to identify current research as belonging to nanotechnology by the visions that researchers publicly propagate" (p. 293).

The third definitional approach he presents is that of "real definition" which "refers to a list of specific research topics." Schummer lists many examples of nanotechnology including microelectronical mechanical systems (MEMSs), nanoparticle research, and targeted drug delivery. "[D]espite its substantial shortcoming of liberally attaching the nanolabel," Schummer opts for this definition "because that is how scientists, science managers, business, and the media mostly use nanotechnology nowadays" (p. 293).

Among scientists and engineers, popular definitions of NS/NT tend to be nominal definitions as in the following definition by Ratner and Ratner (2003, p. 7):

> *Nanoscience* is, at its simplest, the study of the fundamental principles of molecules and structures with at least one dimension roughly between 1 and 100 nanometers. These structures are known, perhaps uncreatively, as *nanostructures. Nanotechnology* is the application of these nanostructures into useful *nanoscale* devices.

Although there is significant disagreement about the definition of "nanotechnology" (with some scientists arguing that the label is simply a new construct for old processes), I will use the broad definition used by the participants in this study (even those who did not approve of the definition used it in this way); that is, that nanotechnology refers to both the tools and the ability to manipulate matter at the nanoscale (one-billionth of a meter). Throughout this text I use the following terms synonymously: nano, nanotech, nanotechnology, nanotechnologies, and NSE. In doing so, I followed the lead of the participants in my study and the literature I encountered. Although I agree with Schummer's 2007 point that nanotechnology is an inaccurate term in the singular since there are presently several hundred different technologies in existence that can be considered nanotechnologies, I found that the singular term was used most frequently by participants in my study and have thus opted to use it throughout the chapters that follow.

SEI Raised by NSE

A key aspect of nanotechnology is that it enables, potentially, the discovery of new phenomena and applications such as those that mimic the powerful capabilities of nature (Alivisatos et al., 2001). Once mastered, the potential that such ability appears to hold is seemingly all-pervasive in improving the quality of life, with developing applications in the manufacturing of improved materials for aircraft and automobiles, faster and smaller communications devices and electronics, more efficient environmental innovations, improved security devices, and targeted drug delivery systems (Alivisatos et al., 2001; Jelinski, 1999). Perhaps one of the most exciting potential applications of nanoscience is the treatment (and even elimination) of disease through nanomedical applications. Some proponents of the technologies even go so far as to predict human immortality made possible by the convergence of genetics, nanotechnologies, and robotics (see, e.g., Kurzweil, 2005).

Such broad and potentially transformative applications have raised many questions among scholars in humanities, the social sciences, and legal studies about the probable, possible, and unknown effects, both beneficial and adverse, of such technologies on society.

CHALLENGES TO SEI RESEARCH

Although the US Federal Government has prioritized research on NSE since the passage of the 21st Century Nanotechnology R&D Act in 2003 through funding large-scale projects such as the National Nanotechnology Initiative (NNI), less than 0.4 percent of the federal funding of NSE research has been allocated to addressing the social and ethical implications of such research (Bennett and Sarewitz, 2006). Thus, challenges to addressing the social and ethical impact of NSE include a 15-year gap between the development of the field and participation by social scientists, limitations on funding, and new and complex scientific and technological concepts (see Bennett and Sarewitz, 2006 for a full discussion of the first point; see Johnson, 2007 for a fuller discussion of the 21st Century Nanotechnology R&D Act).

CONCURRENT FEDERALLY FUNDED PROJECTS ADDRESSING SEI RESEARCH

As the popularity for all things nano continued to increase, so did the funding for projects addressing social and ethical aspects of the technology, and although some scholars argued that the degree of funding was inadequate, particularly for policy-related work on specific applications of nanotechnologies such as weapons development (see Altmann, 2004 for a full discussion of this area), research centers dealing with SEI related to nanotech formed at universities around the United States, the United Kingdom, and the European Union. Some, such as the NNI and the National Nanotechnology Infrastructure Network (NNIN), which are discussed next, were only tangentially concerned with addressing SEI and the rationale for doing so appears to be in the service of facilitating NSE development rather than raising substantive social and ethical questions about the technologies. The remaining projects reviewed later were explicitly concerned with studying the broad social and ethical implications of nanotechnologies.

The National Nanotechnology Initiative. The NNI was the first US Federal initiative to address nanotechnology and nanoscience, and, according to the NNI website, was "established to coordinate the multiagency efforts in nanoscale science, engineering and technology" (NNI, 2007). The rationale for these efforts is described as contributing to "the promise of nanotechnology to improve lives and to contribute to economic growth..." (NNI, 2007). Twenty-six federal agencies participate in the NNI, including the Departments of Defense; Health and Human Services, Food and Drug Administration; Homeland Security; and the NSF.

The National Nanotechnology Infrastructure Network. Launched in 2004, the NNIN is "an integrated partnership" of thirteen nanotechnology user facilities "providing unparalleled opportunities for nanoscience and nanotechnology research" (NNIN, 2008a). The user facilities are located on university campuses across the United States. Eight of the thirteen partnering institutions are involved in research on SEI related to nanotechnology. According to their website, the SEI research in the NNIN serves:

> a dual purpose. First, it constitutes an enabling infrastructure aimed at fostering exchange, producing resources for research and education, and catalyzing additional social and ethical research and discussion. Second, researchers associated with the NNIN will explore issues of ethics, communication, workforce change, industrial innovation, and other social implications of nanotechnology, especially as they emerge in the NNIN community. (NNIN, 2008a)

Additionally, the NNIN identifies the following broad areas of SEI related to nanotechnology:

economic and political implications of potential technology
implications for science and education
medical, environmental, space exploration, and national security implications social, ethical, legal, and cultural implications. (NNIN, 2008b)

From 2004 through 2008, the Center for Workforce Development at the University of Washington was funded through the SEI component of the NNIN. I developed and implemented the study that follows over the 4 years (2005–2008) that I worked as a research assistant on the SEI team.

The NSF funded two Centers for Nanotechnology in Society, one at Arizona State University and the other at the University of California, Santa Barbara for a decade (both centers closed in 2016). The explicit purpose of these centers was to study various aspects of nanotechnology in society.

Finally, a number of university research centers based in the United Kingdom and the European Union focused on various efforts to engage the public in discussions about nanotechnology at the outset of research and development in order to promote ethical development of technologies (see, e.g., DEEPEN, 2007). These examples are just a few of the growing number of federally funded projects popping up at universities around the world in response to calls for consideration of SEI in nanotech R&D.

In addition to the aforementioned federally funded projects on SEI in nanotechnology, there were a handful of privately funded nano SEI research groups as well. The most prominent of these was the Project on Emerging Technologies (PEN) which was established in April 2005 and funded by the Woodrow Wilson International Center for Scholars and the Pew Charitable Trusts. Some of PEN's projects include a US NanoMetro map showing where nano research is highest in the United States, an inventory of nanotechnology-based consumer products, and a series of reports on social and ethical implications of nanotechnologies.

Although most of the aforementioned SEI initiatives are now ended, the professional society for this subject, the Society for the study of Nanoscience and Emerging Technologies (S-NET) formed by science studies scholars from across the United States and Europe in 2009 remains active as does the premier professional journal devoted to the topic, *Nanoethics*, which began in 2007. *An Encyclopedia of Nanoscience and Society* was published in 2010 (Guston).

HISTORICAL ANTECEDENTS

In addition to the relatively recent projects reviewed earlier, it is useful to consider the historical context of federal funding of SEI related to science and technology. The Human Genome Project (HGP), which was begun in 1990 and completed in 2003, marks the first time federal funding of a major research project included allocation for research on ethical, legal, and social issues related to the primary research area (Juengst, 1996). For many scholars HGP also marks the beginning point of scholarly emphasis on these issues. However, as Johnson (2007) points out, the rhetoric around and the funding of these issues in relation to nanotechnology is "unprecedented" in its scope and scale[3] (p. 22) and thus necessitates consideration on its own terms as a scholarly endeavor *sui generis*, rather than just as an another example of federal funding of SEI related to science and technology.

Relevance of Biotechnology Controversy. The public protest and controversy that took place in the United Kingdom in the 1990s over genetically modified food and agriculture, has been described as a model of technology development to be avoided if nanotechnology is to be developed responsibly and receive public support (see, e.g., Kearnes et al., 2006). Torgersen and nineteen colleagues (Torgersen et al., 2002) traced the history of the biotech controversy in Europe from the early 1970s when biotechnology was developed, to the early 1990s when public opposition to "frankenfoods" (a term coined by English professor Paul Lewis to refer to genetically modified foods, see Herrick, 2008, p., fn 2) began to develop, to heightened opposition in the mid-1990s and beyond. Torgersen et al. conclude that the "controversy is still ongoing, and there are few indications that it will soon come to an end" (p. 79).

Although opposition to genetically modified organisms (GMOs) among European publics had characterized the years of 1990–1996, according to Torgersen et al., concerns had been allayed by European Union regulations. However, in 1996 when the United States imported "the first GM food crops" to Europe, "this trigger event led to renewed public and NGO protests" that the authors date from 1996 to 2000 (p. 61) and that continue at the time of this writing. According to Torgersen et al., the GMO debate in Europe was influenced by two contemporaneous life science events: mad cow disease and Dolly, the cloned sheep. The first event took place in 1996 when "the British government conceded, after years of rumour, that there might be a link between human Creutzfeld-Jakob disease and Bovine Spongiform Encephalitis" (p. 62). The second significant event was the cloning of the first sheep that took place in 1997 and led to worldwide debates about the ethical limits of science (p. 64).

Torgersen et al. note that the debate in Europe and the United Kingdom did not extend to the United States and suggest that this was due to distinctive societal issues that differed among the countries. The aforementioned beef crisis in the United Kingdom was one such issue. Another factor that Torgersen et al. point out is that during the same time period of the GMO debate in Europe and the United Kingdom, the US public was involved in a debate over the HGP. The debate over the HGP featured concerns about medical privacy and insurance eligibility, issues that are not relevant to most European healthcare systems (p. 71).

The parallels between biotech and nanotech are many. For example, Torgersen et al. describe biotechnology as deviating from the "traditional" path that a new technology follows when introduced into society, that of society "catch[ing] up" with the new technology (p. 21). The reasons for this change, they suggest, include uncertainty regarding risks, "a difference between public and expert experience of what biotechnology really 'is,'" that biotech is "usually invisible to laypeople," that it takes place "behind closed doors in laboratories indistinguishable from the outside," and that "its products are not at all specific" (p. 22). "For experts, on the other hand," they add, "[biotechnology's] character is no different from that of other technologies" (p. 22). All of the aforementioned points can be made about nanotechnology. Additionally, the authors point to the dual role of European governments as both funder and regulator of biotechnology as generating distrust among European publics who perceived the dual role as a conflict of interest. It remains to be seen whether the American public will perceive as problematic the dual role that the United States government serves as both the primary funder of and regulator of nanotechnology.

Scholarly Context

This project is situated among the array of research from various disciplines addressing the social and ethical implications related to nanotechnologies in particular, and to science and technology in general.

My interest in the discourse of nanotechnology stems from a broader interest in how communication both creates culture and is itself created by culture (see Philipsen, 1987). Specifically, I am interested in how the discourse around nanotechnology creates, or contributes to the maintenance of a scientific culture and how this culture enables or constrains the discourse that represents it. The primary theoretical framework that supports and guides this study is cultural communication theory (Philipsen, 2002) and speech codes theory (SCT) (Philipsen, 1997; Philipsen et al., 2005).

Hymes and the Ethnography of Communication

Dell Hymes combined his academic training in anthropology, folklore, and linguistics to form a new field that he initially called ethnography of speaking (Hymes, 1962) and later called ethnography of communication (Hymes, 1974).

One of the first scholars to answer Hymes' call for ethnographies of speaking, Gerry Philipsen first completed a groundbreaking study of ways of speaking in a south Chicago neighborhood (Philipsen, 1975) and later provided a comparative analysis with an ethnography of mainstream American ways of speaking (Katriel and Philipsen, 1981). Philipsen's proposal of cultural communication (first introduced in Philipsen, 1987) "brought together two important strands of earlier research on culture and communication...(1) difference across groups in terms of communicative practices and (2) the role of communication as a resource in managing discursively the individual-communal dialectic" (Philipsen, 2002, p. 53).

Cultural communication theory is based on two key principles. First, that "[e]very communal conversation bears traces of culturally distinctive means and meanings of communicative conduct" (Philipsen, 2002, p. 53). Philipsen defines "[a] communal conversation" as "a historically situated, ongoing communicative process in which participants in the life of a social world construct, express, and negotiate the terms on which they conduct their lives together" (2002, p. 53). That "members of every group or community partake of a communal conversation" is, according to Philipsen, "a universal aspect of human experience" (2002, p. 55). What is not universal to this experience, however, are the "means" (e.g., "languages...interpretive conventions, ways of speaking, and genres of communication") and "*meanings* of these means," that is, how people interpret the means (2002, p. 55, italics in original).

The second principle of cultural communication theory is that "[c]ommunication is a heuristic and performative resource for performing the cultural function in the lives of individuals and communities" (Philipsen, 2002, p. 59). By "heuristic," Philipsen means that the rules for speaking can be learned by attending to the speech itself. By "performative," he means that communicative conduct can be used purposively to participate in a given communal conversation. The "cultural function" to which he refers in aforementioned Principle 2 earlier is defined as "what a group or individual has settled, or is trying to settle, as to how individuals are to live as members of a community" (2002, p. 59).

Concerning the relationship between speech codes and cultural communication, Philipsen (2002) said:

> A speech code, [SCT] posits, implicates a distinctive way of answering the following questions: What is a person, and how is personhood efficaciously and properly enacted communicatively? What is an ideal state of sociation, and how do people efficaciously and properly link themselves into such states through communicative conduct? What are efficacious and proper means of communication, and what meanings are expressed in and through their situated use? (p. 56)

That the answers to the aforementioned questions differ within and between communal conversations is a central premise of cultural communication theory.

Following the introduction of cultural communication theory in 1987, SCT was introduced in 1992 (Philipsen), formally explicated in 1997 (Philipsen), and revised in 2005 (Philipsen, Coutu, and Covvarubias). The theory comprises six propositions which assert that in a given speech community, one or more cultural codes implying a "distinctive psychology, sociology, and rhetoric" (Philipsen, 1997) can be identified through the communicative conduct of its users. Moreover, the theory asserts that an understanding of such speech codes used enables one to both predict and control some aspects of communicative conduct among those who use the code(s).

According to Philipsen et al. (2005), each of the six propositions that compose SCT "is formulated so as to be amenable...to empirical evaluation...[including]...further substantiation, empirical elaboration, or empirical challenge" (p. 58). It is my intention with the present study to contribute to a fuller understanding of SCT in one or more of the aforementioned ways. I will return to this matter in the final chapter. In what follows I will present each of the six propositions of SCT as laid out in Philipsen et al. 2005 (pp. 58–63), with special emphasis on Proposition 3.

Proposition 1: Wherever there is a distinctive culture, there is to be found a distinctive speech code.

The concept of culture as a code is central to this proposition and to SCT as a whole. Culture in the service of SCT is defined as "a system of symbols, meanings, premises, and rules" (Philipsen et al., 2005, p. 58). The presence of such a system indicates the existence of a speech code, which is "a system of socially-constructed symbols and meanings, premises, and rules, *pertaining to communicative conduct*" (Philipsen, 1997, p. 126, italics added).

Proposition 2: In any given speech community, multiple speech codes are deployed.

Proposition 2 was an additional proposition to the original formulation of SCT that was motivated by Coutu's (2000) and Covarrubias' (2002) research using SCT to identify multiple codes and by extant research on SCT in which the presence of multiple codes is indicated (Philipsen et al., 2005).

Proposition 3: A speech code implicates a culturally distinctive psychology, sociology, and rhetoric.

What Proposition 3 suggests is that talk about talk is not just about talk. According to Philipsen et al., "... wherever and whenever one hears talk about communicative conduct one also hears talk about persons, society, and rhetoric" (2005, p. 61). Psychology in this use refers to how the speakers describe or imply ideas about what it means to be a certain kind of person in light of communication, for example, a man should hit unruly children, not talk to them (Philipsen, 1975), "an 'expert' should speak to the 'average citizen' using 'simple terms'" (Leighter, 2007, p. 201), or a scientist should not talk about ethics, as the present study will illustrate. Sociology refers to ideas about the nature of society, the nature of groups in a society, and the nature of the possibilities or preferred forms of sociation. For example, in some Finnish communities, the preferred form of sociation among groups is one in which the communicative conduct that links people to one another is characterized by quiet observation and respectful silence. This particular speech code values a view of society as one in which its members show respect toward each other by not speaking until they "have something socially worthwhile to say"

(Carbaugh, 2005, p. 123). Rhetoric refers to ideas within a given speech community about what is appropriate and effective communicative conduct. For example, allowing a close friend to speak on your behalf was considered appropriate and effective within the Teamsterville community that Philipsen studied (1975). However, this type of communicative conduct was not valued within a mainstream American speech code and, in fact, devalued as Philipsen describes (1975).

Proposition 3 is the primary proposition of SCT on which I focused my study of scientific ways of speaking about nanotechnology. Specifically, I looked for what was culturally distinctive about the ways in which this speech community understood notions of identity, social relations, and the role of communication. I used a general frame of "talk about nano" (using one of my respondent's words) to look for distinctive and shared ideas among my respondents and the speech community I studied regarding "the nature of persons, social relations, and the role of communicative conduct in linking persons in social relations" (Philipsen et al., 2005, p. 61).

For example, I looked at how respondents used and responded to terms like "nanotechnology" and "social and ethical issues." I selected these terms because they were prominent in the overall discourse about nanotechnology as I indicated in the previous chapter. Following the methodological lead of SCT (see Proposition 5 later), I looked at metacommunicative terms, including "communicate," "explain," and "talk." I also looked at instances of "say" and "tell," and although there were minimal occurrences of this pair, the ones that were present added another layer of understanding to the code as a whole, as Chapter 6 will illustrate. Taking guidance from Proposition 3, I also listened for how the respondents talked about strategic conduct in linking themselves with scientists and nonscientists. I employed Hymes' SPEAKING framework to observe how the respondents treated a variety of speech events, including the interview with me. In Chapters 4–6, I will present and analyze materials from my study that indicate what is culturally distinctive within the scientific speech community I studied about these three areas.

Proposition 4: The significance of speaking is contingent upon the speech codes used by interlocutors to constitute the meanings of communicative acts.

This "interpretive" proposition of SCT (Philipsen et al., 2005, p. 62) says that the meaning of a communicative act will depend on the speech codes the speakers and their listeners are using. Communicative acts or behaviors will be interpreted using the interpreter's speech code. Some examples of these behaviors included silence, speaking, and a full range of verbal and nonverbal behaviors (e.g., whistling, smiling, etc.).

Proposition 5: The terms, rules, and premises of a speech code are inextricably woven into speaking itself.

Proposition 5 provides methodological direction for conducting an analysis of communicative conduct using SCT. Proposition 5 directs the researcher to look for elements of the speech code, that is, "terms, rules, and premises" (Philipsen et al., 2005, p. 62) within the actual speech and communicative conduct of the speakers. Additionally, Philipsen et al. suggest the following for places to look for the speech code: (1) metacommunication, that is, terms used to talk about talk; (2) "rhetorical moments" (p. 62), for example, situations when metacommunicative terms are used; (3) "the contextual pattern of communicative conduct," as within the SPEAKING framework; (4) "and such special forms of communicative conduct as rituals, myths, and social dramas" (p. 62).

Proposition 6: The artful use of a shared speech code is a sufficient condition for predicting, explaining, and controlling the form of discourse about the intelligibility, prudence, and morality of communicative conduct.

This proposition suggests that a given speech code exerts considerable discursive force on those who share it. Moreover, those using the code can do so deliberately to shape the communicative conduct of others who share the code.

SCT thus offers a theoretical resource that can be used to discover and illuminate what is culturally distinctive about the ways in which nanotechnology is talked about within a particular segment of the scientific community, that is, scientists and engineers who self-identify as working in nanotechnology. For the researcher embarking on an investigation of a culture that is not his or her own, a cultural codes approach to communicative conduct is useful in obtaining even a rudimentary understanding of the cultural worlds of the community being investigated.

Additional Theoretical Influences

In addition to SCT, my orientation toward the analysis of the data I collected for this project is informed by a variety of intellectual approaches from a number of disciplines, including sociology (e.g., Bourdieu, 1991) and literary criticism (e.g., Bakhtin, 1981).

Bourdieu's Theories of Language

Pierre Bourdieu conceptualized linguistic competence in economic terms, equating a speaker's degree of linguistic competence with possession of capital. Bourdieu referred to this ability (or lack thereof) as cultural capital that afforded its users with economic and symbolic power (1991). For example, a user's ability to appropriate what Bourdieu called "legitimate" language (1991, p. 97) affords the competent user with access (e.g., to education, employment, social status, etc.) that is denied to those who are unable to effectively appropriate the legitimate language. For Bourdieu, all of social life takes place as part of a larger power struggle between different groups in which more powerful groups use language as a means of control and subjugation of less powerful groups. For their part, less powerful groups take part in this process by recognizing the more powerful linguistic groups as such (Webb et al., 2002). In this way, Bourdieu argued, powerful groups, structures, and language practices are continuously reproduced in a complicit, albeit largely unconscious, coproduction between dominant and dominated groups.

Although Bourdieu talks about legitimate linguistic competence largely in connection with what he calls formal or authoritative discourses such as legal, scientific, medical, academic, or religious discourses, he also acknowledges the legitimacy of what he calls counter discourses. According to Bourdieu, despite the apparent "naturalness" (p. 100) of informal linguistic practices (as opposed to the rules of formal linguistic practices such as legal language), these linguistic (and communicative) practices have their own "rules [and] principles" (p. 100) of use. As an example, he describes the communicative interactions that take place at a neighborhood pub. Indeed, in some settings, such as the pub Bourdieu describes, this informal linguistic practice (or alternative speech code) is the dominant mode of communicative conduct.

In its attention to both the structured nature of speaking and the deliberate and strategic use of a given speech code, SCT acknowledges the "linguistic counter-legitimacy" (Bourdieu, 1991, p. 98), power, and force within a given social sphere, of those speech codes used by less powerful, or dominated, groups. Aside from these limited spheres, or "refuges" (Bourdieu, 1991, p. 98), however, do less powerful speech codes have any real purchasing power (to use Bourdieu's economic metaphor) in the larger marketplace of social life? Moreover, does SCT account for whether or not members of a less powerful speech community (in relation to the larger social sphere) are conscious of the linguistic power dynamics in which they take part and, if so, whether they are capable of changing these power dynamics or not?

Bakhtin's Dialogic Model in Cultural Codes Research

In addition to the ideas about language and power outlined earlier, the present study was influenced by the ideas of Russian literary theorist Mikhail Bakhtin (1895–1975) who described language as multivocal (Bakhtin, reprinted in English in 1981). According to Bakhtin, any given discourse includes multiple voices (even if silent) that are represented by the selection of what is said and not said. Bakhtin also described language as housing ideology. Both of these concepts are relevant to the goals of the present project, namely, to identify what speech codes are present and to elucidate the values these codes embody. The work of Bakhtin has successfully been appropriated by two prominent cultural codes scholars, Philipsen and Katriel, to inform their research and advance the discipline. Before I turn to a discussion of these scholars' work, however, I will briefly review the connections between Bakhtin's ideas and the larger enterprise of ethnography of speaking, of which SCT is a part.

Even without Bakhtin's direct influence, Hymes' early formulation of the ethnography of speaking is consistent with Bakhtin's ideas. For example, Hymes (1962) rejected the idea that "one language = one culture" (p. 131), defining a speech community in what appear to be Bakhtinian terms, as inclusive of all "dialects and languages [that] are in use" (p. 108). By extension, Hymes' argument suggests that multiple voices are present within any given speech community, invoking Bakhtin's concept of heteroglossia.

Bakhtin's notion of dialogism, that "everything...is understood...as part of a greater whole" (1981, p. 426), invokes Kenneth Burke's (1973) concept of synecdoche, a concept utilized by Hymes in his SPEAKING framework. Additionally, Bakhtin's idea that language (and by corollary, any communicative act) is filled with meaning that is context based; that is, that the same utterance (or communicative act) will have a different (and distinctive) meaning depending on its context (i.e., who is speaking, to whom, from what position, for what purpose, in what setting, with what means, etc., see Bakhtin, 1981, p. 401) is directly complementary to

Hymes' approach in studying communicative events as culturally situated. In fact, Bakhtin's discussion of the "actual meaning" of a word being "the actual and always self-interested *use* to which this meaning is put and the way it is expressed by the speaker" (1981, p. 401), mirrors Hymes' assertion that communicative meaning arises from the participants themselves (see Hymes, 1962, p. 119).

Another Bakhtinian concept that is consistent with the ethnography of speaking as conceptualized by Hymes is the notion that language is constructed in social interaction. Bakhtin described the social construction of language in the following way:

> ...the word does not exist in a neutral, impersonal language...it exists in other people's mouths, in other people's contexts, serving other people's intentions: it is from there that one must take the word and make it one's own (1981, p. 294).

This depiction is consistent with Hymes' view of language as socially constructed among a community of participants, a view further developed by Philipsen in his theory of speech codes (1997).

Two of SCT's six propositions are particularly evocative of Bakhtin's notion of heteroglossia and language as bearing ideology. Proposition 2 asserts that "In any given speech community, multiple speech codes are deployed" (Philipsen et al., 2005, p. 11). This is consistent with Bakhtin's concept of heteroglossia, that multiple voices are present within any speech event. Furthermore, Proposition 3 states that "A speech code implicates a culturally distinctive psychology, sociology, and rhetoric" (Philipsen et al., 2005, p. 16), an assertion that is clearly consistent with Bakhtin's description of language as socio-ideological (see Bakhtin, 1981, pp. 271–272).

The social and ethical implications of NS/NT afford one area where an examination of scientific discourse may provide both "descriptive" and "theoretic" value (two goals of the ethnography of speaking per Philipsen and Coutu, 2005, p. 357). An examination of the discourses surrounding NSE would provide descriptive value in identifying the speech codes within a particular scientific "communal conversation" (Philipsen, 2002). Additionally, such analysis has the theoretical value of suggesting how to understand, engage in, challenge, or otherwise appropriate the discourse (Philipsen, 1997, p. 121). This research also has the theoretic value of adding to the cumulative extant literature on sociolinguistics and science and technology studies.

Research Questions

With the aforementioned descriptive and theoretic goals in mind, the overarching research question guiding this study asks what communal conversation is taking place about the values of technology and science in general, and about nanotechnologies in particular, within the scientific community. This purpose is expressed by the following research questions:

RQ 1: How do these scientists and engineers working at the nanoscale talk about their research?

RQ 2: How do these scientists and engineers working at the nanoscale talk about social and ethical implications of nanoscale science and engineering (NSE)?

I began my field activities expecting to find multiple codes at work in the conversation taking place within the scientific community regarding nanotechnologies; however, consistent with Bourdieu's description of language as symbolic power (1991) and Proposition 2 of SCT (Philipsen et al., 2005), I suspected that a dominant code would be present around which alternate codes would circle, attempting to interact but not having the power to do so.

These considerations are addressed with the following research question:

RQ 3: What speech code (or codes) are evident in the communal conversation about NSE among these scientists and engineers?

As stated earlier, scientific discourse is traditionally considered authoritative discourse. As such, does it reflect the less powerful discourses that exist about nanotechnologies and, if so, how? What are the "intentions and accents" (Bakhtin, 1981, p. 293) of NSE discourse in the scientific community? That is, what are the aims of those who use the language of NSE discourse and what are the culturally distinctive (e.g., Bakhtin's "accents") aspects of this language in use? Additionally, what socioideological worlds does the discourse house? In sum, I have attempted to identify and analyze the discourses of nanoscience in order to provide a way "to guess at and grasp for a world behind their mutually reflecting aspects" (Bakhtin, 1981, p. 414) and, in so doing, addressed the aforementioned research questions.

Description of Materials, Data, and Methods

I used a combination of ethnographic approaches and discourse analysis to collect and analyze the data for this study. Modes of inquiry included fieldwork, participant observation, and interviews.

Description of Materials

Data analyzed for this study were obtained from two main sets of materials. The first set of materials I collected was based on my fieldnotes from two years (2005–2007) of experiences in nanoscience classrooms, attending scientific lectures, participating and observing three conferences (one on nanoethics, one on nanotechnology, and one on technology in general), and conducting informal interviews with various stakeholders in these arenas. My materials also included twenty in-depth interviews that I conducted with scientists at the University of Washington (UW) associated with the UW Center for Nanotechnology (CNT) in which they discussed their research and their views of SEI posed by the research.

Fieldwork. Over the 4-year (2004–2008) course of this project, I conducted fieldwork by attending three conferences (two in the State of Washington, one in the State of South Carolina) focused either exclusively or in part on nanotechnology development or ethics of nanotechnology; attending six lectures on NSE (or related topics such as science and society) by visiting scholars to the University Washington and to the City of Seattle; enrolling and participating as a graduate student in two courses within the nanotechnology program at the University of Washington; and by observing three two-hour training sessions for new users at the UW Nanotech User Training Facility. In addition to the readings that were assigned for my classes on the subject, I read as much as I could find on the subject of NSE from a wide range of sources. This included dozens of policy-level reports, scholarly journal articles, and press releases (which I received from a variety of listservs I subscribed to, including both industry and academic perspectives on developing NSE); several science fiction titles that were recommended to me by my interviewees (e.g., Michael Crichton's "Prey" and Isaac Asimov's "I, Robot"), and recent bestsellers either promoting or discouraging development of NSE (e.g., Ray Kurzweil's "The Singularity is Near" for the former perspective, Bill McKibben's "Enough" for the latter perspective).

My involvement as a participant in the NNIN afforded me, to some degree, an insider's perspective. I was privy to meetings, e-mails, and other organizational documents that I would not have had easy access to were it not for my role as a participant. I include as part of my fieldwork the many conversations I had informally with graduate students studying nanotechnology and with faculty members. These were not formal or structured interviews, but instead natural interactions that took place over the course of the 3 years that I was researching this project. The conversations took place over lunch, after class, or during group work in class. In all cases, I told the interactants about my project and that I was trying to understand more about the subject and how scientists thought about their work in relation to social and ethical obligations. I believe that these interactions provided me with a valuable perspective from people who could be considered cultural informants. I used the data I obtained from these field-level activities as a repertoire of experiences that helped me to (a) orient to the specific community I was investigating, (b) develop an appropriate methodology (as described later) to investigate this community, and (c) provide a lens with which to assess the interview data I collected.

Interviews. In addition to my fieldwork activities, I conducted twenty semi-structured interviews with scientists and engineers working at the nanoscale. My single criterion for inclusion of an interviewee in this study was that the interviewee be affiliated with the UW's CNT. At the time of the study, the CNT included eighty-one faculty members from the Departments of Chemistry, Physics, Bioengineering, Chemical Engineering, Electrical Engineering, Materials Science and Engineering, Biochemistry, Genome Sciences, Physiology and Biophysics, and Microbiology. The interviews were completed by telephone (nine) and in person (eleven). I transcribed the interviews in full after I completed them. The interviews ranged from 15 minutes to 2 hours and resulted in more than 200 single-spaced, typed pages of transcribed data. My final corpus of this first set of materials consisted of my fieldnotes from my ethnographic work and twenty transcribed interviews, thirteen with male faculty members and seven with female faculty members.

Data Analysis

Briggs' final phase of conducting fieldwork using the interview involves analysis of the collected materials. When the analysis was completed, he advised, the researcher should look at the interview responses more closely in light of what he or she has learned about the culture as a whole through the research process described earlier. I analyzed the data I collected from the interviews using discourse analysis to identify and explicate participant meanings. Additionally, I considered these responses in light of what I had learned during my field-level activities. Although I consulted the fieldnotes I had made from these activities, I did not conduct a formal analysis of these data, opting instead to use the data set as a background for understanding the interview data and indicating (or supporting) the use of particular speech codes. Using an approach described by Schatzman and Strauss (1973, p. 117), I conducted multiple readings of all of my materials, moving back and forth between my fieldnotes, my observational notes, and the interview transcripts to complete my analysis.

Briggs advocated a two-step process in the analysis of interview data. First, he recommended looking at each interview "as an interactional whole" (p. 103) using a heuristic aid in the analysis of transcripts and observational data. He used a form of Hymes' SPEAKING[4] framework (1972), a dialectical system in which one or more of the elements of a communicative interaction can be examined in relation to each other. In order to identify and interpret the significant components of the interview, Briggs recommended taking detailed notes directly after each interview and then considering the linear structure of the interview (e.g., the format). I followed these recommendations in my analysis, taking notes during each interview, afterward, and during the transcription to orient myself to the overall experience of the interaction. Additionally, I used Hymes' SPEAKING framework to consider significant elements of each interaction.

The second step that Briggs called for in his two-step analysis process is that of "interpreting individual utterances" (p. 105). In order to understand an interviewee's individual answers, Briggs suggested the analyst pay close attention to the metacommunicative features of the interaction, considering visual, prosodic, and verbal cues. Secondly, he recommended that the analyst check her interpretations of the interviewee's responses by carefully considering ongoing context that was created by both interview and interviewee during the interaction.

Thus I approached the analysis of the interview data in the following way. First, I printed out copies of each transcribed interview and read through each one on its own terms, making notes in the margins and underlining words and phrases that caught my attention. I also began to map out the linear path of the event, as Briggs recommended, noting how the interview began, how it proceeded, and how it ended, and whether there were any shifts in key or genre.

Second, I analyzed the individual responses I received to my questions from each interviewee, considering the metacommunicative features of each response and verifying my interpretations by considering the context as Briggs advised. In all, I read each transcript in its entirety a minimum of six times and reviewed those portions I highlighted during these readings numerous more times throughout the analysis and writing up phases of the project findings.

Presentation of Results

I have included relevant sections of the transcripts in the chapters that follow. In presenting portions of the transcripts, I have used a convention illustrated in Johnstone's 2002 text "Discourse Analysis." Thus, I number each line of text, allowing the breaks in text to occur at the right margin, and including in italics a reference number to indicate the respondent. The transcription conventions I used are from those presented in Silverman's 2003 text (see p. 254). I have also reproduced those conventions in Appendix A. For ease in reading, I have numbered each excerpt included sequentially.

Notes

1. Japanese engineer Norio Taniguchi is credited with being the first to use the term, which appeared in a keynote conference paper in 1974 to refer to tools for characterizing nanoscale materials; however, the term did not receive widespread attention until Drexler's usage in the 1980s.
2. The atomic force microscope enabled scientists to image, measure, and manipulate matter at the nanoscale.
3. At a 2007 conference for philosophy and technology, one scholar remarked that he had become interested in nanotechnology after a colleague encouraged him to "feed at the nanotechnology trough."
4. SPEAKING is an acronym representing various components of a communicative interaction: setting, participants, ends, act, key, instrumentalities, norms, genre. This list is meant to be exhaustive and as such, not all elements were addressed by Briggs in his analysis, nor will be addressed in the present analysis.

CHAPTER 2

Discovering a Code of Scientific Communicative Conduct

Abstract Findings are presented from a 4-year-long ethnography of communication conducted within the community of nanotechnology research at the UWCNT. Some understandings of a scientist's nature, or psychology, are presented, suggesting that the way in which a scientist is taught to think of himself or herself as a scientist does not allow for open consideration of SEI broadly defined.

Keywords Ethnography of communication · Nanotechnology · Social and ethical issues · Speech codes theory

ORIENTATION TO RESULTS CHAPTERS

In this chapter and the two that follow, I consider findings from my analysis of a 4-year-long ethnography of communication I conducted within the community of nanotechnology research at the UW. In particular, I address my first three research questions: (1) How do these scientists and engineers working at the nanoscale talk about their research? (2) How do these scientists and engineers working at the nanoscale talk about social and ethical implications of NSE? (3) What speech code (or codes) are evident in the communal conversation about NSE among these scientists and engineers. In regard to the third research question, I will limit my discussion here primarily to the dominant speech code I observed in the speech community.

Consistent with ethnographies of speaking, this study was situated in the professional, organizational life of the participants and members of this speech community, that is, scientists and engineers working in nanotechnology. It was largely limited in scope to one "segment of a given speech community's resources for communicative conduct" (Philipsen and Coutu, 2005, p. 359), language in use among scientists and engineers in a given social situation, namely, an interview. The goal of my research was to identify "the means and meanings" (Hymes, 1972) of speaking within this community in order to ascertain what one needs to know about communicative conduct to participate effectively in this speech community. The focus of this type of research is on observation of the speech act, the perspective of the participants regarding what is appropriate communicative conduct, the "whole spectrum" of events and means possible in the given community (Philipsen and Coutu, 2005, p. 358), and treating this spectrum as a system by integrating "language use and social situation" (Philipsen and Coutu, 2005, p. 358).

In order to discover the "system" of means of communication available in this community, I observed as many speech events as possible within it, including speech acts in class, at conferences and lectures, in group conversations, and in dyadic interviews (Philipsen and Coutu, p. 358). In so doing, I considered as many speech events and means as possible to form a "system" of means. Although I focus primarily here on one segment of this speech community's means (see Philipsen and Coutu, p. 359), talking to an outsider about their work, this segment is part of a larger body of data that included additional speech events. Rather than forming the subject of my study, this means of speaking is simply a context in which I found materials and data that helped me identify a speech code deployed within this community, both in talking to an outsider about their work and in other settings.

The Speech Community

The subject of this ethnography of communication is the speech community or "social unit" (characterized by Philipsen and Coutu as "the starting point of the ethnographic study of communication in a given case," 2005, p. 367) that comprises the scientists and engineers associated with the CNT at the UW. The speech community I investigated also includes students who are studying with the scientists and engineers. Early on, as I conducted fieldwork and analyzed my materials, I noticed an important distinction between science and engineering, as well as between "basic" and "applied" scientific research, as highlighted by the respondents themselves, that I originally thought might warrant treating these two areas as separate speech communities. For example, respondents referred to significant differences in language among the various science and engineering disciplines associated with the Center for Nanotechnology in the following ways (the numbers in parentheses following the quotes refer to the participant interview from which the quote is excerpted):

> "I mean it's hard talking the same language" (0069)
> "people use different vocabularies in different disciplines" (0068)
> "every discipline has slightly different languages" (0067)
> "I don't think there are that many of us that...can speak the same language" (0062)
> "a lot of the time just goes into explaining each others' languages" (0057).

After analysis of the data, I concluded that there was sufficient overlap to justify considering these various disciplines within the scientific community holistically, as co-members of a larger speech community, particularly since they appear to use the same speech code regarding understanding of scientists and engineers, the appropriate role of the scientist or engineer vis-à-vis the public, and the various functions of communication as strategic conduct (i.e., the distinctive psychology, sociology, and rhetoric that the code implicates). Additionally, despite the difficulties with different terminologies used among the disciplines that respondents reported, they also reported being able to overcome these difficulties with time. One respondent said that such language differences are "prevalent in science" (0063). Thus, it appears that language differences notwithstanding, the respondents are able to understand each other sufficiently to engage in interdisciplinary research. The ability to do so suggests a shared code of communicative conduct. Throughout the following analysis, therefore, I use the term "scientist" generally to refer to all of the respondents I interviewed, except for sections where I make a deliberate distinction between scientists and engineers.

Discovering the Code

When I began this study with the intention of discovering a code or codes of communicative conduct at work within the scientific community at the UW, I was worried about the possibility of not finding a code, specifically I wondered if a code even existed. I came to realize, however, that such a worry was unfounded and illogical. Scientists and engineers, like any other group of people, depend on a shared set of understandings in order to communicate with each other about their particular task at hand, that is, developing scientific knowledge and technological applications. Moreover, according to the first proposition of speech codes theory, "[w]herever there is a distinctive culture, there is to be found a distinctive speech code" (Philipsen et al., 2005, p. 58). Philipsen et al. use the term "culture" to refer to "a code" that "consist[s] of a system of symbols, meanings, premises, and rules.... about many aspects of life" (p. 58). The aspect of a given culture to which a speech code refers is that of the "symbols, meanings, premises, and rules" that refer to communicative conduct. These codes have been historically, and are continually, created by members of the speech communities in which the codes are deployed.

Speech codes theory also indicates explicitly where to find evidence of a speech code. According to Proposition 5 of speech codes theory, "The terms, rules, and premises of a speech code are inextricably woven into speaking itself" (Philipsen et al., 2005, p. 21). One respondent in particular referred to a shared way of communicating among scientists as a "code" with the expression "we know our code" (0070), when speaking about strategic use of language. Such references served to reinforce the validity of the idea that there is indeed a speech code in use among the members of this speech community.

Philipsen, Coutu, and Covarrubias also referred to "rhetorical moments" (2005, p. 22) as opportunities for observing a speech code in use. I observed one such moment during a question and answer session at a panel that took place at a nanoethics conference. A presenter had suggested that anti-dystopian views of thinkers such as Orwell and Huxley might be instructive in considering SEI related to nanotechnologies. After his presentation, he was admonished by Mihail Roco, who is perhaps the leading figure in US federal funding initiatives of nanotechnology, that this conference was a scientific conference and that the presenter should not return until he had something scientific to say. This outburst by Roco suggests several premises and norms regarding scientific culture: that discussion of philosophical issues is not scientific and that a conference devoted to nanoethics is not the appropriate place to discuss philosophical issues, including, ironically, ethics, as "ethics" is largely understood by philosophers. I will return to a consideration of how ethics is understood by members of the scientific community. The rhetorical moment described earlier suggests that a violation of some element of a scientific speech code had occurred and thus indicated that there was indeed such a code.

In the report of my analysis that follows I will attend largely to Proposition 3 of speech codes theory: "A speech code implicates a culturally distinctive psychology, sociology, and rhetoric" (Philipsen et al., 2005, p. 61). In the present chapter, I will consider aspects of the code that indicate what is culturally distinctive in this speech community regarding understandings of human nature (psychology). In Chapter 4, I will consider understandings of social relations (sociology), and in Chapter 5, I will consider understandings within this speech community regarding strategic conduct (rhetoric) (Philipsen et al., 2005, p. 61). I will consider each of the three areas by attending to key symbols and their meanings, premises, and rules that correspond to each of the three inter-related areas. I attended to two elements in particular from Hymes' (1972) speaking framework, "participants" and "ends/goals" to help elucidate these elements.

The Psychology of a Scientist

During a graduate nanotechnology class in which I was a student, I often asked the members of the small group to which I was assigned what their views were on SEI, particularly as these issues were not addressed in our weekly readings that the class discussed in small groups. My questions were generally met with uncomfortable silence and I recall feeling that my questions were inappropriate but not understanding why. One of my group members kindly informed me at one point that the technologies we were reading about and discussing were "being developed for good" (0081).

This type of resistance, even refusal, to discuss negative implications of scientific work was observed throughout my fieldwork and interviews and for the early months of my data collection. I felt frustrated by this reaction until I realized that it signaled something significant about the way members of this speech community understand themselves and their work. In particular, I began to understand that the scientists I encountered during the course of my fieldwork understood talking about broad SEI as being inconsistent with their understandings of themselves as scientists. As one of my small group members half-jokingly put it, "scientists aren't supposed to worry about ethics" (0087), a statement I understood to be both definitional and normative.

With this in mind, I began to ask myself why it was that "scientists aren't supposed to worry about ethics." In what follows I present a series of excerpts from my larger corpus of fieldwork and interviews that I have selected to address questions about the psychology of a scientist. From these excerpts, I construct datasets of terms that are indicative of how a scientist in the speech community in question talks about himself/herself. I then offer several cultural premises and rules that I suggest are part of the speech code in use in this community. I close with presentation of additional materials that suggest additional ways scientists talk about themselves that deviate from the official scientific discourse exemplified in elements of the code presented previously.

What SCT Means by "Psychology"

In what follows, I use the term "psychology" in a particular way that is consistent with how the term is treated in speech codes theory, as "implicat[ing] a view of what a person is and how persons are constituted" (Philipsen, 1997, p. 138). In other words, "psychology" in the service of speech codes theory refers to "a local code about the nature of persons" (Philipsen, personal communication, 2008). Thus, when I use the term psychology here, I am not talking about how the members of the speech community I investigated *think*, but *their notion of how people think*. My aim is to determine what the scientific speech code I discovered says about the nature of persons from the standpoint of those who use this speech code.

Beginning Fieldwork

After spending nearly a year reading about nanotechnology, the exciting breakthroughs in medicine, robotics, and textiles projected in National Science Foundation reports, as well as the many concerns associated with these breakthroughs that were raised by researchers in the social sciences, I began my fieldwork phase. I did so expecting to readily encounter what scientists actually working in nanotechnology had to say about the alleged "promises and perils" (a term used by Kurzweil, 2005, among others writing on the subject) of nanotechnology. Instead, I was surprised to find no discussion of this type taking place within the scientific community and found myself coming up against a brick wall of resistance when I inquired about scientists' perspectives on SEI related to nanotechnologies.

I began my fieldwork activities by attending nanotech seminars open to the public at the UW, listening intently for some reference to the issues that had been raised in my literature review, but heard nothing aside from rather dry technical descriptions of the given speaker's research. During these early months in the field I also attended a nanotechnology conference featuring industry and university researchers working in nanotechnology. As I attended each session and the subsequent question and answer sessions, I detected some tension around the term "nanotechnology" and began to suspect that discussing SEI was not culturally appropriate in this speech community. I had previously detected both of these areas as problematic, both the term itself (nanotechnology) and the topic of SEI related to nanotechnology, particularly in an early interview with a senior nanotech faculty member in which he had discussed the likelihood that additional faculty members would identify social and ethical concerns with nanotechnology. The following excerpt was preceded by a question about whether scientists would perceive there to be risks associated with research in nanotechnology:

Excerpt 1

1 *(0066)* I don't think that, you know, that there's from the scientists- I don't
2 think you're gonna get people saying that they're, that they have big
3 concerns, you know. That they, uh, will undoubtedly tell you that there
4 should be proper, you know, regulation and it's gonna be kind of an
5 evolution to figure out how to set up proper regulatory agencies to keep
6 track of these things, you know. But the innate fear of, uh, I mean, there's
7 no big difference in my mind between nano (7.0) technology and
8 chemistry all along, you know. It's just another chemical, you know. (xxx)
9 everything is another chemical that you have to worry about. It's no
10 different when you get devices than microelectronics or biomedical

11	devices, you know. There's a certain amount of, uh, regulation to be done
12	as these things evolve, you know. I don't see it as being fundamentally,
13	uh, different anyway. All that stuff (xxx) will ultimately be handled by the
14	same <u>regulatory</u> things that are already covering- mostly chemical type
15	stuff.... you have to develop new regulatory mechanisms as new dangers
16	come up (in this) you know, eh, uh, and I think scientists will be very
17	supportive of doing that, um, but they also are resistant to sort of knee jerk
18	reactions that are extreme reactions, to perceived dangers that aren't really
19	there.

Several aspects of the aforementioned excerpt were particularly helpful in my early formulation of the appropriate ways of communicating within this speech community. First, the respondent clearly suggests that scientists are not concerned about risks associated with nanotechnology. He references these concerns in the following ways, in this excerpt and elsewhere in the interview:

"[no] big concerns"
"the innate fear of [nano]"
"new dangers"
"knee jerk reactions"
"extreme reactions"
"perceived dangers"

In all of the earlier instances, except for the reference to "new dangers," the language used suggests an unscientific perspective, "reactions" that are based on "fear" or "perceived dangers" rather than reason or fact. The statement that scientists do not have "big concerns" suggests that scientists are in control of their research; that is, they might have "small" concerns, but nothing that they are unable to handle. This confidence in scientists being able to handle whatever may arise is reinforced by the respondent's reference to scientists being "very supportive" of addressing "new dangers" (lines 15–17) that may arise. These references effectively diminish the possibility of adverse implications of nanotech research, implicitly suggesting that to think otherwise would be unscientific, because, after all, scientists themselves are not worried about these issues. This usage is also suggestive of an attitude that scientists are simply not concerned because there is nothing about which to be concerned.

Second, in line 7, the respondent had one very long pause of seven seconds between "nano" and "technology" (the pause was an anomaly as there were not additional pauses of this length in the remainder of the interview), suggesting a discomfort with using the term in this way (as opposed to "nanoscience"). Earlier, in fact, the respondent had stated that scientists did not consider themselves to be working in nanotechnology, thus differentiating between theory and application and highlighting the ambivalence many basic researchers feel toward identifying their work as "nanotechnology."

Finally, the co-occurrence of the terms "regulation" or "regulatory" with the terms "evolution" or "evolve" bears noticing. Out of the four uses of the term "regulation" or "regulatory," the respondent uses the word "evolve" or "evolution" twice to describe the process of adjusting regulations to meet the needs of nanotech research and development:

> "there should be proper, you know, **regulation**" (lines 3–4)
> "it's gonna be kind of an **evolution** to figure out how to set up proper **regulatory** agencies" (lines 4–5)
> "There's a certain amount of, uh, **regulation** to be done as these things **evolve**, you know" (lines 11–12)
> "All that stuff (xxx) will ultimately be handled by the same **regulatory** things that are already covering- mostly chemical type stuff" (lines 13–15).

The aforementioned usage suggests inevitability in the process of nanotech development and implies that "proper" regulation (which he refers to twice in lines 4 and 5) will automatically occur. In his last usage of the "regulatory" in line 14, the respondent emphasized this word, suggesting its importance to his statement. I refer to this aspect of the code of science I observed in this speech community as a deterministic stance toward the development of science and technology and will return to this theme at a later point.

During a question and answer session at a nanotechology conference presentation that took place at the UW, it was clear that discussing SEI related to science and technology was an inappropriate topic for basic researchers. During the presentation, the presenter had said that the goals of his research were to "emulate proven biological designs" and "impart life-qualities to robust engineering materials" (0088). After the presentation, an audience member asked that, given such fascinating goals as emulating nature, what kinds of applications might such research have and what kinds of issues might they raise. The presenter answered, "Applications? We don't really think too much about applications? We're just chemists" (0088). He laughed somewhat nervously and his answer was met with scattered laughter throughout the audience and a frustrated shrug from the questioner. At this point, another presenter on the panel spoke up and said that it would require at least 15 years to get through the US Food and Drug Administration approval process to market a biomedical application, suggesting that the path from research to product development was a lengthy one involving many different agents and implying that it was quite understandable that the presenter would not have an application in mind.

The then director of the UWCNT said that the term "nanotechnology" was problematic because it did not appear to include nanoscience when, in fact, research termed "nanotechnology" encompassed such areas as nanoscience and biotechnology. This comment seemed to imply that perhaps the questioner had misunderstood the intent of the research presented which, although termed nanotechnology was in fact, nanoscience. From this exchange, I understood that those doing basic research at the nanoscale (nanoscience) did not find discussions of application to be relevant to their work, an assumption that was later verified in many conversations I had with scientists and engineers both in informal conversations and in the formal interviews. (The comment also indicated that there might be particular motivations for terming one's research "nanotechnology" and I will return to a discussion of these motivations in subsequent chapters.)

Given the aforementioned experiences, I began to formulate the following observations regarding this speech community: (1) that the term nanotechnology was perceived as problematic within this community although they all used it to describe their different, often completely unrelated, areas of research in both basic and applied contexts; (2) that those identifying themselves as doing basic research did not want to discuss potential applications of their work and, in fact, doing so seemed to run counter to their identity as basic researchers (a point to which I will orient in this chapter); and (3) that although engineers had no problem discussing intended applications of their research (e.g., faster and more precise gene-mapping technology), they were unwilling to discuss long-term or wide-scale implications of their research (e.g., What are the social and ethical implications of gene-mapping?) and seemed genuinely perplexed by my questions about broad social and ethical implications of their work.

In the previous chapter (p. 67), I noted that scientists and engineers described the research they conducted in their different disciplines (e.g., chemistry, engineering) as nanotechnology. In the present chapter, I will orient to the second assumption identified earlier, that for those identifying themselves as doing basic research, discussing potential applications of their work appeared to run counter to their identity as basic researchers. Regarding my third finding, that engineers seemed genuinely perplexed by my questions about broad social and ethical implications of their work, I offer the following example in which the engineer I interviewed reacted with considerable discomfort to my question about whether there were SEI related to his work in nanoscale instrumentation:

Excerpt 2

1 (0071) Yeah, not the stuff we do, I mean, you know, there uh, there is certainly,
2 uh yeah, when you're looking at human cells that's already available, you
3 know somebody has taken, I don't think, uh, we even have to go through
4 human subjects even, because it's considered available samples.... So, uh,
5 existing samples. Now if you're going to take new samples, that would be
6 different. If you're going to study the disease then of course you have to
7 worry about consent issues and so on... And that's certainly there. But
8 otherwise I don't see for our research any particular harm (hhh).

After the interview from which the aforementioned excerpt was taken, I made the following observational note in my fieldnotes: *He seemed really nervous talking to me.* (Notes for Interview 0071). I based the aforementioned observations on the fact that his comments in Excerpt 2 were characterized by excessive (compared to other parts of his interview) disfluencies and false starts. Additionally, I noted during my analysis of this interview that the respondent's laughter at the end of the comment reproduced earlier did not seem consistent with the subject matter. Thus, I interpreted the disfluencies, false starts, and laughter as nervousness on the part of this respondent.

As the months went by and my time in the field increased, I had conversations with graduate students in science and engineering that helped me better understand the site of my research. During one conversation with a group of graduate students from different disciplines who shared a common research focus on nanotechnology, the students told me that they could hardly understand many of the speakers who visited the campus to give the public nanotech lectures. They explained that each discipline had a slightly different set of terms for talking about the same phenomena. Another student told me that there was a hierarchy of sorts within the sciences with physics occupying the top role, that of pure science, and the others falling underneath depending on their emphasis on application, with engineering being an example of an applied field. These conversations helped me understand that there were many different types of researchers within the nanotech community. As other scholars have observed, often the only thing uniting these researchers is the label of "nanotechnology" itself (e.g., Schummer, 2004). Was there, I wondered, something more that united these researchers? Specifically, I wondered if there was a shared code of communicative conduct from which both scientists and engineers from a variety of disciplines drew, a code regarding the nature of persons? I set out to discover the answer to this question.

As I began my formal interviews with nanotech researchers from across disciplines, I heard repeated emphases on the differences between basic and applied science, particularly from those doing basic research. Such comments were generally offered as a reason why the respondent did not consider SEI of nanotechnology of concern in their research. The three responses that were obtained from my audio-recorded interviews are presented later and are representative of the types of responses I heard numerous times throughout my 4 years in the field:

Excerpt 3

1   *(0056)* ...my research is at the pretty basic side so I'm not on the applied side so
2   it [regulation of nanotechnology] hasn't really impacted me.

Excerpt 4

1   *(0057)* I mean, these [social and ethical issues] are larger issues and, you know,
2   we are too, I mean, very much at the scientific end of this that we don't
3   really deal with that, you know, immediately in our work.

Excerpt 5

1   *(0064)* ...right now I'm so deep into what I call basic, basic, basic science that
2   I'm very far away from this [knowledge about funding sources for
3   nanotech research].

What the aforementioned excerpts have in common is that the speakers rationalize their lack of familiarity with issues of regulation, broad SEI, and funding sources for nanotechnology by identifying themselves as doing basic research:

"at the pretty basic side"
"very much at the scientific end"
"deep into basic, basic, basic science"

The implication is clear: those doing basic science do not "deal with" the "larger issues" of social and ethical implications of science and technology, to which the aforementioned respondents referred as regulation of nanotechnology, broad social and ethical implications, and funding. I was thus able to determine that within this speech community, discussions of social and ethical implications were associated with applications and since applications were not relevant to basic research, neither were discussions of SEI. But what about for those doing applied work,

I wondered. The following responses were received from three respondents doing applied research when I asked for their perspectives on SEI related to nanotechnology:

Excerpt 6

1 (0059) Yeah. So, yeah, I heard about these a lot too and, uh, I think they're for
2 people focusing on applications in biomedicine?...so for myself, because
3 my application focus [is] more on standard electrical engineering
4 applications...which is not, you know, science related...I'm much less
5 worried about any kind of, uh, ethical or social issue.

Excerpt 7

1 (0061) I think there are a lot of societal impacts...I don't necessarily think my
2 own research [in computing] is gonna be the make or break on it.

Excerpt 8

1 (0064) Because we do computer work we don't run into the same ethical issues as
2 you might get in other areas....to me personally, it [SEIN] doesn't mean
3 anything...maybe it <u>should</u>, but if you take it to other <u>design</u> work or say
4 you're trying to use, um, artificial materials and implant them, you know,
5 in humans...yeah, there are ethical issues.

From these materials, I surmised that among those doing applied work, they considered only biomedical applications to have relevance for consideration of SEI. The next logical place to look for answers, I decided, was from biomedical researchers themselves. What follows are the responses referencing SEI that I received from three respondents working in biomedical applications:

Excerpt 9

1 *(0060)* [what is needed is] some routine toxicology...protecting our students and
2   researchers at the moment....

Excerpt 10

1 *(0072)* The ethical consideration is if we don't apply a new set of tests and FDA
2   approval requirements on a nanoparticle as we did to its components, we
3   have a problem.

Excerpt 11

1 *(0075)* ...there's no nanorobots running around...that is not the risk, it's more
2   from a health risk of what are—what if these particles go in our body,
3   right, and people work with them, like in a workforce, that's the first
4   important thing, right, they have these particles in the air um, in high
5   concentrations, um, are they dangerous?

The following set of terms drawn from the aforementioned excerpts illustrates how those respondents working in biomedical applications expressed the SEI associated with these applications:

> "protecting our students and researchers"
> "apply[ing] a new set of tests"
> "[applying a new set of] FDA approval requirements"
> "[determining] the health risk[s]"
> According to the excerpts from which the aforementioned terms were taken, such actions as "protecting," "applying," and "determining" are
> "[what is needed]"
> "The ethical consideration"
> "the first important thing"

The aforementioned terms indicate that those researchers working in biomedical applications who cited concerns that they had about these processes and products were primarily concerned about ensuring that the appropriate tests and regulatory processes were followed to ensure product, and worker, safety.

Across my collected materials, "social issues," when defined, were understood by the respondents as health and safety issues and were talked about by those I interviewed to be relevant to them on a local scale only and not unique to nanotechnology. That is, the health risks were described as the respondent did in Excerpt 9: "[what is needed is] some routine toxicology...protecting our students and researchers at the moment...." (0060). The use of the phrase "routine toxicology" implies that this is standard practice throughout science, not unique to nanotechnology. The emphasis on "students and researchers" suggests that this is the locus of concern, rather than the public at large.

In the following excerpt, the respondent says that his work with individual nanostructures is "about the least dangerous thing I could imagine doing":

Excerpt 12

1 *(0076)* You look around you and see other people who are—pretty much any
2 other experiment, I couldn't see as actually more dangerous than what I'm
3 doing. Just studying something very tiny, an inert, on a bench, in my lab,
4 is about the least dangerous thing I could imagine doing. Much more
5 dangerous being in the kitchen, or crossing the road.

I received some version of the aforementioned response in the numerous informal and formal conversations and encounters I had with members of the nano community during the course of this project. I repeatedly heard that nanotechnology did not raise SEI that warranted any more concern than any other area of science, and may warrant even less concern, as the following excerpt from one of my recorded interviews indicates:

Excerpt 13

1 *(0060)* Why just nano? There's a lot more dangerous–in fact if you ask me,
2 there's nothing to the alleged quote, unquote, dangers of nano. There's not

3   a single thing there, it's totally fallacious. Nano is the safest thing we're
4   doing. There's a lot of other things that are less safe.

The relative safety of individual nanostructures notwithstanding, the respondent in Excerpt 12 went on to say that ensuring lab safety was always a priority:

Excerpt 14
1 *(0076)* And you do continuously have to—you continuously face situations
2   where you do have to assess the risk. And so, it's not that you can
3   completely put it out of your mind. We're faced with health and
4   safety issues in the lab and you do have to think about these all the
5   time.

Like the respondent earlier, those respondents who gave examples of social issues related to nanotechnologies primarily discussed the safety of their labs. One respondent said that he "avoided working with materials that have toxic effects" (0074). One engineer told me that researchers in her lab group wore double gloves when working with nanoparticles. Of course, she added, they know that such a precaution is not helpful because nanoparticles are small enough to pass through the gloves, so her preferred way of handling the uncertainty was "not to think about it" (0067). A graduate student in the sciences told me that she managed such uncertainty by reminding herself that a certain amount of risk is implied in doing science.

Similarly, across the interviews, specific concerns about nanotechnology were articulated by the respondents by addressing (1) issues of privacy threats raised by pervasive computing and surveillance technologies:

"ubiquitous computing" (0061)
"where I see the risk is really in terms of... this privacy issue, can somebody plant
something on me" (0058)
and (2) health issues related to nanoparticle exposure and environmental impacts:
"anti-bacterial particles... [in] the waste stream forever" (0061)
"I wonder... does it [carbon nanotubes] make any fibers that when we breathe

them in we get cancer more often" (0062)
"people don't quite know what these things [nanoparticles] are going to do to
them in the long-term" (0057)

Thus, the respondents did identify specific concerns they, as scientists, had about nanotechnology, largely framing these concerns in questions about long-term human health and environmental effects. These concerns notwithstanding, the respondents who offered the examples such as earlier, qualified their comments with follow-up statements such as the following:

Excerpt 15

1	*(0062)*	Definitely important to think about, but not, not any more alarming than
2		any of those other things [e.g., how to dispose of old computer monitors].

Excerpt 16

1	*(0056)*	[nanotech is] no different than any other scientific project.... those
2		[questions about long-term impact] are questions that get asked not only in
3		nanotechnology but in lots of different things.

Excerpt 17

1	*(0058)*	as a scientist, we can make a fair assessment of total danger and say this is
2		low or moderate or this can have tremendous implications and and, uh,
3		you know, uh, do our assessment and our protection based upon, you
4		know, sort of rational assessment of how dangerous is this really.... with
5		some good study and weighing the benefits and looking at where the
6		products [are] ending up and uh actually being proactive about this, I don't
7		see such a tremendous risk.

Placing the aforementioned excerpts into the SPEAKING framework provides a clearer picture of the participants, the acts they describe, and the tone they use to discuss their perspectives on risks of nanotech. Although Excerpt 17 is the only one of the three excerpts to explicitly name the participant using the term "scientist" (line 1), it can be assumed that, as scientists being interviewed as scientists, the other two speakers were speaking as scientists. Thus, the views they offered were "as a scientist" (line 1). The acts attributed to scientists in the aforementioned excerpts include six active verbs:

"think about"
"[ask] questions"
"make a fair assessment"
"[do] some good study"
"[weigh] the benefits"
"[look] at where the products [are] ending up"
"[be] proactive"

Taken together, these terms describe a picture of a scientist in his or her lab, thinking hard, asking questions, making assessments, weighing benefits and risks, looking at the life cycle of nanoparticles, doing good studies, and so on. It is a reassuring image of a rational, serious, objective, and well-trained authority figure. The tone of the aforementioned excerpts is further indicated by the following set of descriptors:

"not...alarming"
"no different [from other science]"
"[not] a tremendous risk"
"rational assessment"

Three of the four expressions aforementioned deal with minimizing concern about nanotech risk. The fourth indicates why the lack of concern: not because scientists do not care but because they are accustomed to doing "rational assessment of [dangers]" and will continue to do so with nanotech as with "any other scientific project" (0056).

Taken together, these statements suggest that scientists are capable of making the appropriate judgments and decisions about nanotechnology, just like they do with "any other scientific project," and manage the unknown risks associated with them. These sorts of qualifications helped me understand that scientists have faith in science and in themselves as scientists. They do, in fact, think from time to time about big picture issues associated with their research and quickly determine that, "as a scientist," they are capable of handling them. After all, thinking about these issues is "the best we can do" (0076), as one of the respondents said.

An interview that took place during the latter phase of my data collection was particularly helpful in explicating the psychology of a scientist, particularly regarding SEI. This respondent was a scientist who had extensive experience working with ethicists. In the excerpt that follows, she explains that thinking about SEI runs counter to a scientist's self-perception:

Excerpt 18

1 *(0073)* I think scientists in general are always interested in bettering society so the
2 idea that anything they could do could harm society is not a part of-cuz
3 they spend countless hours trying to do things that would help people, not
4 to hurt people, so um, but thinking about all the ways your research <u>could</u>
5 hurt people is important...

As this statement suggests, among members of this speech community, even *thinking* about "the ways your research could hurt people" (lines 4–5) is not a part of the nature of the scientist, who by the very definition is someone who is "always interested in bettering society" (line 1).

It is important to note that I went into this project with my own ideas about what constituted "social and ethical implications" of science and technology, specifically, consideration of broad, normative questions of ethics, social justice, and responsibility. What I found expressed in the speech community of the engineers and scientists I interviewed was the assumption that these issues are implicit within the larger goal of science, that is, knowledge making and scientific discovery. Thus, my questions about whether and how such issues were considered were met with somewhat annoyed bemusement by some respondents such as a scientist who described his view as "basic 101 science ethics":

Excerpt 19

1 *(0076)* You just have to have general principles that you apply to all of science and um take bits of science, case by case.
2 *(DB)* Yeah. What would be an example, just, of a general principle.
3 *(0076)* Oh, OK. Well, now you're asking me really difficult questions.
4 *(DB)* Oh, really? (hhh).
5 *(0076)* OK, I guess the general, basic general principle is that you, you should
6 look at the, you should be thinking about the possible applications of
7 what you're doing while you're doing it. And the possible dangers
8 involved as well as the possible benefits. And that's essentially, is the
9 essence of the ethics of the business, whatever you're doing, the best you
10 can do is not only look for look for the positive, but also look at the
11 negatives. It's the best we can do.

The respondents did not seem readily able or willing to provide examples of "SEI" related to their work in nanotechnology. In the aforementioned excerpt, for example, the respondent gave a general answer to my question about how scientists address broad SEI in their research. When I asked him for more explication, he said I was asking him "really difficult questions" (line 3), which was somewhat ironic considering he had already characterized his ideas on the subject as "basic 101 science ethics." He then suggested that "thinking about the possible applications" (line 6) of one's work in science was "the best we can do" (line 11) to address SEI.

As I mentioned in the previous chapter, although I suspected early on that the phrase "social and ethical issues" was not something that is intelligible or considered important in this speech community, I persisted in using it because the concept was central to the goals of this project, namely to understand how, if at all, members of this speech community understand SEI related to their work. Thus, when the participants asked me to provide examples of what I meant by "social and ethical issues" I gave vague, general answers and explained that I was really interested in how *they* defined this term, if indeed it was a meaningful or useful term to them.

As presented earlier, respondents who talked about social issues understood these issues primarily in terms of the health and safety of the scientists doing the work, rather than in terms of the larger effect of the work on others. When defined, ethical issues were defined by the respondents as standard scientific practices and examples were given of transparency in data collection, use of informed consent procedures, and avoiding plagiarism in publication. For example, one respondent made the following utterances in reference to what he perceived as ethical issues: "be honest," "disclose...how you took the data and what assumptions you made," and, "the scientific give and take of you question somebody else's data and they question yours and you go back and forth and, you know, develop hypotheses" (0056). Another respondent summed up the general response I had received regarding ethical issues related to nanotechnologies in the following way:

Excerpt 20

1 *(0058)* I would think that people don't really don't what the ethical implications
2 are going to be–the ethical implication of plagiarizing or making up data,
3 now they are quite aware of that...

Of the twenty respondents, only one raised issues of long-term, broad social effects and normative ethical questions, a respondent who identified herself as having extensive experience working with ethicists and acknowledging that her views were not the norm among scientists. Indeed, when I probed further with the other respondents, the standard response I received was that such issues are not their domain as scientists, as the following excerpt indicates:

Excerpt 21

1 *(0057)* I mean the social and ethical issues are, I mean they are, they are
2 meaningful in the larger context, yes, but most scientists, I mean most

3	academic scientists in universities, are dealing with things on a very local
4	scale...s:::o, unless they have something that is <u>really</u> at the point of
5	field testing and things like that, they don't really deal with it.

As Philipsen has observed, "Not every meaning is given a word in every language or culture" (1992). Overall, "social and ethical issues" appears to be a term that is largely without use in this speech community and apart from a value
about carefulness within and among scientific workers for their own health and on-the-job personal safety that is implicitly understood. Moreover, within the corpus of speech I heard in this study, I did not encounter a term addressing the study of long-term and widespread social impact or of broad, normative ethical questions.

The following excerpt is an example of the type of response I received when I asked about SEI related to nanotechnologies:

Excerpt 22

1	*(0056)* Well, I think that it's no different than any other scientific project, you
2	have to pursue it in an ethical and (hh), you know, and socially responsible
3	manner, right? I think it's, you know, you want to be honest in what you
4	do and you gotta disclose the, you know, how you took the data and what
5	assumptions you made and, uh, you know, and and there's the scientific
6	give and take of you question somebody else's data and they question
7	yours and you go back and forth and, you know, develop hypotheses and,
8	uh, you know—to me it's the same, you know, you need to follow the
9	same sort of ethical and social guidelines that you do in your—you know,
10	that we've been doing in research for years.

Four things are worth pointing out here. First, the suggestion that addressing SEI related to nanotechnologies is no different than addressing SEI in any other scientific project:

> I think that it's no different than any other scientific project, you have to pursue it in an ethical and (hh), you know, and socially responsible manner, right?

Additionally, the respondent's use of the verb modal "have to" implies that scientists are required to consider SEI in their work, *all* their work, not just nanotechnology.

Second, and more importantly, this behavior is already, and has been "for years" (line 10) taking place in the scientific community:

> to me it's the same, you know, you need to follow the same sort of ethical and social guidelines that...we've been doing in research for years.

The respondent emphasizes twice that attending to "the same sort of ethical and social guidelines" is "the same" behavior that has been taking place "in research for years."

Third, I found particularly interesting the interviewee's definition of behaving in an "ethical and...socially responsible manner" (lines 2–3) in lines 3 through 7:

> "...you want to be honest in what you do and you gotta disclose...how you took the data and what assumptions you made and...there's the scientific give and take of you question somebody else's data and they question yours and you go back and forth and...develop hypotheses...."

According to this definition, addressing SEI related to scientific and technological development sounds very similar to the scientific method: transparency in data collection, scientific rigor, hypotheses development (see, e.g., Campbell, 1952).

Finally, similar to reactions I received from others I talked to as I described earlier, this respondent appeared clearly uncomfortable answering my questions about SEI. He spoke easily and with few disfluencies until I asked him about social and ethical implications of nanotechnologies. The shift in topic was characterized by a marked increase in his speech of disfluencies including (1) false starts:

"I think it's, you know, you want to be honest in what you do . . .
(2) trailing off:
"you go back and forth and, you know, develop hypotheses and, uh, . . . "
(3) laughter:
"you have to pursue it in an ethical and (hh), you know, and socially responsible manner, right?"

and (4) an excessive (compared to the rest of his speech) use of the expression "you know," which occurs seven times in the aforementioned excerpt. These four indicators taken together suggest considerable discomfort with the topic being discussed.

In the excerpt that follows, the respondent suggests why scientists are not worried about social and ethical implications of nanotechnology or any area of science and technology. The excerpt took place near the end of the interview. The respondent has talked for nearly half an hour without interruption about nanotechnology. During his discussion of scientists' perspectives on nanotechnology, he mentioned the book "The Making of the Atomic Bomb" (1986) by Richard Rhodes, which is a book about the physicists who developed the science that led to the development of nuclear weapons. In the following excerpt he refers to the book again:

Excerpt 23

1 (0065) But I mean one of the, the reasons I thought about that is that one of the
2 things that comes up in that is that, you know, scientists are scientists
3 because they believe that they're working to help mankind by being
4 scientists and they wouldn't do it if they weren't thinking that that was the
5 case. And they also believe in science, they believe in technology. They
6 believe that it will win out and that, uh, no matter what it is we do we will
7 always be smart enough to figure out and we're moral enough in our
8 general behavior to figure out how to, um, make the integrated result to

9	the benefit of mankind and not to the detriment of mankind and, uh, that's
10	just a scientist's attitude and you will get that answer on average from
11	the people that you talk to that are scientists by far I'm sure...

As he predicted, I did receive some version of the aforementioned answer in nearly all of the interviews and fieldwork I conducted, a further example of which is illustrated in the following excerpt:

Excerpt 24

1	*(0074)* ((Clears throat)) I don't see any ethical issues, this is, we are really
2	pushing the limits of uh science and technology, of course it has
3	societal impact. After all, we are engineers and scientists and we are trying
4	to ease the problems that are present in the society. Our materials, our
5	protocols, and our gadgets, new devices, so that is what we are
6	trying to do. So I have never, ever, in fact, thought about this will be
7	becoming an ethical issue.

The notion that scientists are, qua scientists, attuned to SEI is implicit in the aforementioned characterizations of being a scientist. Indeed, part of the nature of scientists as the previous respondent indicated in Excerpt 8: "we will always be smart enough...and we're moral enough...to figure out how to...make the integrated result to the benefit of mankind..." (lines 7–9).

An additional understanding of a scientist's nature was articulated as wanting to climb a mountain simply because it is there. For example, in the following excerpt, the scientist describes science and technology as "a mountain" that scientists "wanna climb":

Excerpt 25

1	*(0073)* So, technology is always, always much more advanced than our- we
2	often tend to (3) think more advanced technology and not think

3   about the ramifications of it. It's just, it's like a mountain–we wanna
4   climb it.
5 *(DB)*   Yeah (hhh).
6 *(0073)* No, really, it's very similar, science is very similar to that.

When I chuckled (line 5), the scientist responded in a serious tone, making it clear that she was not joking with her use of the analogy (line 6).

As the respondent in Excerpt 24 had asserted, she believes that scientists consider their purpose, and the goal of science and technology, to be motivated by a desire to benefit society. Consistent with her assertions, other respondents described the motivations of scientists and engineers in the following ways:

"scientists aren't supposed to worry about ethics" (0087)
"scientists are always interested in bettering society" (0073)
"[scientists] spend countless hours trying to do things that...help people" (0073)
"[as a scientist] you should be thinking about the possible applications of what you're doing..." (0076)
"[as a scientist] you do have to think about [health and safety issues] all the time" (0076)
"people [including scientists and engineers] don't really know what the ethical implications are going to be" (0058)
"[scientists and engineers] are trying to ease the problems that are present in ...society" (0074)
"[a scientist's goal is] to understand things" (0076)
"[scientists and engineers] are pushing the limits of science and technology" (0074)
"[scientists want] to do what you can in general to make this a better place to live" (0060)
"[science is] like a mountain—we wanna climb it" (0073)
scientists "don't really deal with [social and ethical issues]" (0057)
considering SEIN is "no different than any other scientific project" (0056)
"scientists are scientists because they believe that they're working to help mankind by being scientists" (0056)

These characterizations paint a picture of the scientist as someone who is motivated by curiosity, intellectual challenge and the desire to benefit society, what I have termed the altruistic nature of a scientist.

In the section that follows, I will discuss cultural premises and rules regarding the nature of a scientist that are indicated by the data I have presented in this chapter.

CULTURAL PREMISES AND RULES REGARDING THE NATURE OF A SCIENTIST

Considering the aforementioned analyses, an associated cultural premise could be articulated as follows:

> *(P1) A scientist is motivated by the desire to benefit society.*
> *(P2) A scientist is motivated by the discovery of knowledge.*

Similarly, an associated cultural rule could be articulated as follows:

> *(R1) A scientist should not talk about broad social and ethical issues related to science.*
> *Alternative Understandings of a Scientist's Nature*

In contrast to what I will call the altruistic nature of scientists described earlier, some of the respondents I interviewed offered understandings of the psychology of scientists that presented a more faceted, more human, view of the scientist's nature.

Excerpt 25

1	*(0058)* Now in terms of faculty, a lot of people, you know, also want to do stuff,
2	uh, get their stuff published, get recognition, because this the way the
3	system works and, you know, you don't really worry about the ethical
4	implication because you're not actually going to mass produce it, you see
5	what I mean? ... And so, you know, uh, in a way people look at things
6	through a different fashion. I think that that scientists look at things as
7	like, well, what is giving me visibility in my field and recognition and, uh,
8	you know, more grant money.

Excerpt 26

1 *(0076)* I love lots of things about what I do. Um, if I had to pick one thing, I'd say
2 it's the prospect of seeing something fundamental that nobody else has
3 ever seen and and understanding it. And going down in history as having
4 discovered it. That would be... cool... I can't pretend I'm not motivated
5 by the idea that I might become well-known for discovering something...
6 But it's fun doing it even if you don't become well known.

In the next excerpt, speaking specifically about funding and research outputs, the respondent describes scientists as people who do not like being controlled:

Excerpt 27

1 *(0075)* ...we are in a environment in which there is extreme freedom, the
2 academic freedom, the freedom in general, we are not easy to control. We
3 are not really—we don't see a, we don't see ourselves as employees, right,
4 we more see ourselves as, um, freelance, something. And and so it's [a]
5 little harder to get, to control, um, um, the people here at the university,
6 right, so, because they, it's generally not people that like control.

Drawing from the aforementioned materials, the following terms are used to describe the nature of a scientist:

"get their stuff published"
"get recognition"
"what is giving me... more grant money"
"seeing something... that nobody else has ever seen"

"understanding it"
"going down in history as having discovered it"
"we don't see ourselves as employees"
"[we] see ourselves as freelance"
"[scientists are] not people that like control"

The aforementioned depictions of a scientist's nature suggest that a scientist is often someone who does not see himself as having to answer to someone, is motivated by recognition, funding, discovery and understanding of something new, and fame. This understanding of the nature of a scientist provides a more *human* nature of a scientist than the previous depictions do.

ADDITIONAL INSIGHT

During the last six months of my time in the field, I became involved in a network of scholars interested in science studies on the UW campus. The network included basic scientists and I was particularly interested to hear what they had to say. During a panel presentation of scientists discussing philosophical approaches to science, one of the scientists somewhat jokingly described science studies as "dangerous stuff" (0089), explaining that he had "lost a little bit of religion" (0089) after studying social science approaches to science. He added that scientists needed help understanding "which mountain" (0089) they should climb, echoing the reference to science as a mountain that one of my respondents had earlier made but complicating it by suggesting that issues of social and ethical significance should guide decisions about which areas of scientific research should be pursued. Another scientist on the panel, a geneticist, said that "the authority of science in our society has a sort of bulldozer effect" (0090), adding that he did not see any reason that that practice would not continue into the future. A physicist in the network referred to himself as a "rebel" (0083) for addressing broad SEI related to emerging science and technology in his work.

Toward the end of my data collection, I spoke with several social scientists who had either worked with or were presently working with large federally funded projects related to emerging science and technologies. The personal experiences they shared with me confirmed many of the assumptions I had made as I was beginning my analysis of the data. For example, they described their efforts to integrate broad ethical reflection in the research and development of science and technologies to be at odds with a dominant culture of objectivity within science, which one ethicist described as "assumptions of objectivity that almost are part of the identity of scientists" (0077). This adherence to objectivity seems to discourage scientists from thinking about issues they perceive to be "soft and subjective," according to this respondent, such as ethical and social issues, and makes it difficult for them to understand "that, in fact, there's a huge amount of social assumption that goes into what they do or how they think about questions" (0077). Another ethicist I spoke to said that there is reliance among scientists on the mythical understanding of science as "objective and pure" (0078) that results in a disconnection between scientific research and applications of science that will affect the public.

During the course of my fieldwork I spoke with a doctoral student who had left the nanotechnology doctoral program because she became disillusioned when she realized that science was not "pure" (0079) as she had originally believed. She told me that she was disgusted by the way nanotechnology research was focused on obtaining funding and that there was no consideration of social justice or normative ethics in the field. She said that her personal convictions compelled her to change fields and pursue scientific research with an explicit environmental emphasis. As a scientist with these views, she said, she felt that she was "definitely going against the tide" (0079).

Conclusion

In this chapter I have presented some understandings of a scientist's nature based on my fieldwork and interviews. I have suggested that the scientist's proper perception of himself or herself does not allow for open consideration of SEI broadly defined. I will return to a fuller analysis of the implications this finding has for communication in a later chapter. Despite the official understanding of a scientist's nature that I have presented here, many respondents revealed that they were indeed concerned about broad issues related to nanotechnology, primarily in health, safety, and privacy terms. However, they allayed their concerns and uncertainty by relying on the confidence they had in science to address any concerns that arose, thus illustrating the force of the code in controlling the way they thought about these issues. Scientists who openly embraced broad concerns and uncertainty regarding the beneficial nature of science described themselves as at odds with mainstream scientific culture, as "a rebel," "going against the tide," or having "lost a little bit of religion." These various understandings of the nature of science, and the scientist, suggest at least two speech codes within the larger speech community. Additionally, references to science as "having a bulldozer effect" and as "religion" imply that science exerts a powerful cultural force, that I am suggesting is extended to the ways of speaking among scientists. In the next chapter, I will consider understandings of appropriate social relations within the scientific community I studied.

CHAPTER 3

Ways of Speaking about Social Relations

Abstract In this chapter, aspects of the speech code within the community studied that indicate culturally distinctive understandings of social relations are examined. The data builds a multifaceted picture of what the respondents had to say about the following areas of social relations: the different roles for science and industry in society at large, the different roles for basic and applied research, and the role of the media and education in public understanding about science.

Keywords Ethnography of communication · Nanotechnology · Social and ethical issues · Speech codes theory

Overview

In "Speaking Culturally" (1992), Philipsen contrasted two cultural codes of communicative conduct represented by two distinct American cultures he referred to as Teamsterville and Nacirema. For the users of the Teamsterville code, where a concept of honor was paramount, society was understood as "more important than any individual" (p. 15). For Nacirema code users, for whom the concept of dignity was central, society was "of value only in the degree to which it enhances the individual" (p. 16). According to Philipsen, speech codes are resources with which "individuals and societies...answer questions about why they exist and where they fit in a scheme of sense and meaning" (p. 16). Returning to a consideration of Proposition 3 of speech codes theory, a given speech code includes specific understandings of, and ways of talking about, society and social relations (Philipsen et al., 2005).

In this chapter, I will consider aspects of the speech code within the community of nanoscience I studied that indicate culturally distinctive understandings of social relations. I will attempt to answer questions about what the speech code that I encountered in the setting of the nanoscience community at the UW says about social relations and the role of the scientist and science in society. I will interrogate the materials I amassed and analyzed to answer why, according to this speech code, scientists "exist and where they fit in a scheme of sense and meaning" (Philipsen, 1992, p. 16). Science is produced in a social world and, as the materials that follow will illustrate, the scientists I encountered in this community were acutely aware of this.

Perhaps because I initially interpreted responses from those I interviewed implying that there were no social and ethical implications related to nanotechnology as instances of dismissiveness or denial, rather than as expressions grounded in a code of communicative conduct, I had begun to wonder if the stereotype of the scientist cut off from society, alone in his lab, was not so inaccurate after all. However, about half way through my interviews, I began to notice consistent emphasis from the respondents on society and social relations such as public expectations, funding pressures, media portrayals, and public support. From the perspectives of the scientists and engineers whom I came into contact with during the course of this project, their work was highly connected to the social world around them, including the federal government, industry, news and popular media, and the community at large. In what follows, I will present materials from my fieldwork and interviews as well as from secondary sources such as newspaper articles (see Schatzman and Strauss, 1973 for a discussion of the usefulness of such sources in field research) that I collected and analyzed to determine what the code of nanoscience I studied had to say about social relations.

The Roles for Science and Industry in Society at Large

In this section, I present materials from my interviews and fieldwork that indicate different roles for science and for industry pertaining to nanotechnology (and science and technology in general) within society. In the first excerpt I present here, the respondent talks about the need for ethics training in the sciences, how he would handle a hypothetical ethical scenario, and the different roles of scientists and industry related to ethics and research and development:

Excerpt 1

1   (0058) I think that bringing these people [scientists] to think in a different manner
2   <u>does</u> require a proactive type of effort where these people that are training
3   in ethics actually kind of can bring to bear their insights, and make people
4   think differently.... I don't think it's [ethics] ignored though, I think that
5   people, you know, if I were going-let me put it this way-going to
6   discover something that I think would have potentially very, very, ugly
7   application, let's say in terms of bioterrorism or what have you, I mean,
8   you know, you would th-I would think twice about publishing this stuff
9   or even doing something about it.... But, you know, but what you can do?
10 What can you do, because you can sit on something and ultimately
11 everybody has access to the same amount of information, and the ideas are
12 ripe and they will come up one way or another.... And so, you know,
13 in a way people look at things through a different fashion. I think that
14 scientists look at things as like well, what is giving me visibility in my
15 field and recognition and more grant money. What industry looks at is like
16 well, what is gonna turn a profit quickly and how am I gonna cover my
17 butt so I don't get sued.

Three aspects of the aforementioned excerpt are particularly relevant to the present discussion. First, the respondent's statement in lines 1–4 that "bringing [scientists] to think in a different manner <u>does</u> require a proactive...effort" indicates a perceived need within this community, at least from this respondent's perspective, for outside training in how to "think differently" about ethics related to nanotechnologies.

In the previous chapter, I presented the ways of understanding ethics related to nanotechnologies within the speech community of nanotech researchers. The interview and fieldwork data I collected indicated that understandings of ethics were implicit with science and largely limited in scope to "internal ethics" (0089) such as plagiarism, lab safety, etc. Thus, to "think differently" about ethics (line 4) implies thinking about these issues in terms that are different from the present understanding of ethics as internal issues.

The need for thinking differently about ethics was raised by scientists during a panel presentation for social studies of science faculty at the UW. One of the scientists said that although scientists do not need help with science, they could use some "guidance" in ethics (0089). He clarified by saying that he was not talking about "internal ethics" such as fraudulent data but "external ethics" or "what science *should* be done," an area he said that scientists "don't normally think about" (0089).

The second aspect of Excerpt 1 that is relevant to a discussion of social relations within science is seen in lines 4–12 where the respondent says that ethics is not "ignored" in the sciences and as an example says that he "would think twice" before publishing something that might have "very ugly application." He goes on to imply however, that doing so would be futile because "ultimately everybody has access to the same amount of information, and the ideas are ripe and they will come up one way or another" (line 12).

This scenario was invoked by many of the individuals I talked to during my fieldwork activities although, with the aforementioned exception, not raised explicitly in the interviews. I encountered it explicitly during a graduate lecture for nanotech students in which the choices considered for handling broad social and ethical questions were: (1) Let someone else worry about societal implications re: my discoveries, (2) If I don't do it, someone with lower ethical standards might, and (3) I see potential for misuse, so I won't do it (0091). Only the first two choices were considered acceptable by the students during the discussion.

Thus, I understood this perspective, as a taken-for-granted assumption among the members of this community: "let someone else worry about societal implications re: my discoveries because if I don't do it, someone with lower ethical standards might." I also confirmed this assumption in informal conversations with several colleagues in related fields. When faced with the three choices aforementioned, I would have to agree with the students in the class and my colleagues: the third option is generally unacceptable since it can be argued that any research could be misused. However, I was struck by the fact that among the list of possible responses raised during the class that a scientist might choose from to confront SEI in his or her work, there was not a choice that involved ongoing consideration of broad SEI in the research and development of science and technology, perhaps with the collaboration of an ethicist. In other words, the choices considered seemed unduly limited in scope.

As I noted at the close of the previous chapter, there appears to be a deficit in understanding among members of this speech community regarding how to integrate elements of an alternative code that would allow for consideration of broad SEI into the dominant code. Again, I am suggesting that the worldview represented by the dominant code precludes the ability to adequately consider broad SEI because the language does not allow for the possibility and, in fact, actively resists it. Consideration of broad SEI is seen as incompatible with scientific progress.

The final aspect of Excerpt 1, and one which I touched upon briefly at the close of the previous chapter on a scientist's identity, occurs in lines 12–17 where the respondent describes the different motivations of scientists and industry. Scientists, he suggests, are motivated by "visibility," "recognition," and "more grant money"; where industry is motivated by "turn[ing] a profit quickly" and not "get[ing] sued." This implies a related understanding of the roles of each within science. That is, a scientist's role, these data would suggest, is to pursue activities that promote his or her research. These activities include visibility, recognition, and funding. Those in industry would be expected to be motivated by making a profit as quickly as possible, but also would be expected to want to avoid lawsuits. Thus, they can be expected to "cover [their] butt" in order to avoid lawsuits.

The implication here within the larger context of a discussion on broad SEI is that when these issues refer to the safety of the consumer, it is not only the responsibility, but the motivation of industry to ensure that their products are safe. This understanding was reiterated in various ways throughout the interviews: that consideration of SEI (where these are understood as health and safety issues) fall within the domain of "commercialization" (0056) as the following references from the interviews indicate:

Excerpt 2

1	(0056)	...the public's concerns need to be addressed and data shown.... I think
2		there is, when you talk about actually bringing some things to
3		commercialization.... it has to be addressed before a lot of this nano stuff
4		can be dealt with or commercialized.... if you're a company trying to
5		commercialize something and there's mass hysteria in the (hhh) general
6		public, that's probably not a good time to commercialize things.

Excerpt 3

1	(0057)	...when they get to the point of field testing and so then, obviously, they
2		are duty <u>bound</u> to look at social and ethical issues.

Excerpt 4

1	(0067)	I think it's the individual company's responsibility to make sure that their
2		products are safe, I think it will be their moral responsibility.

Excerpt 5

1	(0072)	...the people who really have to pay attention to this [social and ethical
2		issues] is the businesses and people who are in industry more than people
3		who are in academia...

It is clear from the aforementioned excerpts that for these respondents, it is

> a company trying to commercialize something
> they
> the individual company

the people who are in industry
who are
   duty bound
   [have a] moral responsibility
   have to pay attention
in order to
   [address] the public's concerns
   [show] data
   look at social and ethical issues
   make sure that their products are safe

Thus, within this speech community, ensuring product safety is the appropriate role of industry, not scientists. Not only is it the industries' "moral responsibility" (0067) but as the respondent in Excerpt 1 pointed out, those in industry can be expected to ensure product safety if for no other reason than to avoid being sued.

Rather than focusing on product safety, which is, after all, the role of industry according to my respondents, a more appropriate social role for scientists according to this speech code might be phrased, in the words of the respondents, as follows:

discovering new things and understanding new things (0076)

always with the eye that hopefully...you are going to help society in some way (0063)

[in order to] make life as good as possible at a biological level for everybody in the world (0064).

In the next section, I illustrate this distinctive understanding of the role of the scientist in society by presenting interview excerpts and additional examples from my fieldwork activities. I begin with a discussion of tension between basic and applied research that was highlighted in the interviews and fieldwork activities.

Different Roles for Basic and Applied Research

When I began my fieldwork activities, I did not make a distinction between nanotech researchers conducting basic research and those conducting applied research. Early on, however, I realized that the scientists I encountered who described themselves as doing basic research interpreted questions about application as inappropriate and even offensive. For example, after a doctoral student in physics gave a presentation in one of the two nanotechnology courses I completed while I was conducting my fieldwork, I approached him after class hoping to learn more about his research. Afterward, I wrote the following in my fieldnotes:

> Asked him afterwards about application. He faltered a bit but said it was a "fair question" but a "cruel question to ask of physics" since there was never a clear value to the research. Am wondering if asking questions about application is a violation of cultural norms... (Fieldnote entry dated 10/19/05).

I was surprised when the student said that my question was a "cruel question" as I had not asked it in a spirit of cruelty but of genuine curiosity. Moreover, the student quickly excused himself from our conversation and I was left with the uncomfortable feeling that I had unwittingly insulted him.

As the months went by, I gradually came to understand that there is a larger discourse taking place regarding what one scientist I interviewed described as a global push for applied research that had left those doing basic science struggling for funding. In late 2007, for example, US federal budget appropriations resulted in "nearly a half-billion cut" in the US Department of Energy's Office of Science, "the basic research arm of the agency," in a move that was described as "favor[ing] short-term development, which often comes at the expense of long-term research" (Bullis, 2008).

Similarly, government budget cuts in the United Kingdom in late 2007 were projected to "result in a 25% cut in the number of grants the [Science and Technology Facilities Council] can award for research projects" to physics departments across the United Kingdom, a situation *The Guardian* characterized as "the worst funding crisis in more than 20 years" for UK university physics departments (Gilbert, 2007).

Although many of the scientists doing basic research interviewed for this study cited funding woes, many of the scientists and engineers I interviewed who were conducting applied work said that funding was not a problem for them. One respondent said that she did not anticipate any funding obstacles to her work in applied nanotech: "as far as funding's concerned, I don't think that barrier exists, because it's one of the new trendy areas to work in.... it's definitely a well-funded area" (0067). Similarly, another respondent conducting applied nano-related research said that "this is just a wonderful time for receiving funds" (0074). These data appear to support the scientist quoted earlier who claimed that federal funding favored applied research over basic research.

The tension between applied research and basic research that I described earlier appeared to stem from external societal perceptions regarding the value of each and subsequent availability of funding. However, within the community of scientists and engineers interviewed, there appeared to be a shared understanding that each had different roles in society but that both roles were equally important. For example, one respondent said that it was important to have both basic and applied research in nanotech, adding: "I think that's part of the beauty of science is that you have a lot of different people contributing in very fundamental ways with different perspectives" (0069).

In another interview the respondent, a self-described "conventional" scientist, discussed what she perceived to be a fundamental difference between "conventional scientists" and "nanotech people or engineers":

Excerpt 6

1   (0063) [T]here is a step that nanotech people or engineers take that I think
2   conventional scientists don't, most don't, and that as a conventional
3   scientist you're trying to understand this problem from every single
4   direction, every, every which way you can and you're not gonna take that
5   next step to apply it or do something with this until you thoroughly,
6   thoroughly understand it. And there's sort of this break point where
7   engineers will say "I don't understand it, I don't care to understand it," or
8   "it'd be nice to understand it but I wanna go use it and make something of
9   it" and they take that leap and they do all kinds of things that I would say
10 pessimistically, "Well, that won't work, there's no way and you don't
11 know this, this, this," and then they turn around and do it.

The aforementioned explanation was helpful for me in understanding why the basic researchers to whom I spoke were not generally disposed to discuss applications of their research. In lines 2–6, the respondent says that a basic scientist is "not gonna take that next step to apply [their research]...until [they] thoroughly, thoroughly understand it." This characterization suggests clearly why a scientist conducting basic research would not consider it appropriate to talk about applied aspects of their work. In terms of social relations, doing so is the domain of applied scientists and engineers, not basic scientists, whose role it is to "[try] to understand this problem from every single direction" (lines 3–4).

This understanding of basic science is consistent with a definition of "pure science" found in a classic text first published in 1921 entitled "What is Science?" (1952) by the British physicist Norman Campbell:

> ...the motive of our study is supposed to be intellectual curiosity without any ulterior end; and...our criterion will be always the satisfaction of our intellectual needs and not the interests of practical life. (p. 4)

In the following excerpt, the respondent echoed this understanding of the motive of basic scientists; however, he emphasized that such research does indeed serve a practical purpose. During our interview, the respondent had said that as a basic scientist, "I do have a problem when people come ask me what devices I'm trying to make" (0076). Later, I asked if he ever thought about possible applications of his research, expecting him to say "no." Instead, he answered, "Well, of *course*" and proceeded to explain:

Excerpt 7

1 *(0076)* I mean, I can't imagine anybody, if they notice something, if they discover
2 something really interesting, that could be very useful...and they <u>realize</u> it
3 could be useful, not <u>wanting</u> to put it to use.... So, I mean, I think we're
4 all the same in that sense. But some of us find ((exhales loudly)), some of
5 us just have a different philosophy. Well, I don't know, it takes all types,

6	OK. You need to have scientists and you need to have engineers, you need
7	to do basic research and you need to do applied, focused research as
8	well.... you just have to look at the history of society, technology, to see
9	that we wouldn't be anywhere without having people who didn't care
10	at all about applications, just wanted to understand things. You just have
11	to look at the <u>skepticism</u> that met the people who discovered electricity
12	and electromagnetism, about *what use can this possibly be?* ((speaks
13	in an exaggerated, stern stage whisper)) That wasn't what these people
14	were interested in. They were just trying to understand things and history
15	teaches that... discovering new things and understanding new things
16	ultimately has bigger benefits to society than fiddling around just
17	trying to develop things for... certain applications based on already
18	well-established science.

Several statements in the aforementioned excerpt help illustrate further the perceptions among this speech community regarding social roles. First, the respondent says that scientists and engineers are "all the same" in their desire to put their discoveries "to use" (lines 1–4), indicating a shared role. However, he describes a distinct role for basic scientists, as "people who [don't] care at all about applications, just [want] to understand things" (lines 9–10), echoing Campbell's definition of "pure science" quoted earlier (1952, p. 4).

Second, the respondent refers to a historical response of skepticism regarding the practical value of basic research (lines 10–12), and responds to this skepticism by emphasizing that basic science "ultimately has bigger benefits to society" than applied science (lines 15–16). This reference invokes again the larger discourse external to this speech community about the value of basic research to which I referred earlier.

Considering the aforementioned selection of materials in particular aided me in formulating the following premises regarding social relations as understood in this speech code:

> (P3) Both applied and basic perform an important role in improving society.
> (P4) Basic science has not been valued by society as much as applied science.
> (P5) Basic science should be recognized as valuable by society.

Later in this chapter I will suggest what, according to this speech community, are appropriate ways of recognizing the value of scientific research.

Continuing a discussion of appropriate social relations, I now turn to an examination of an interview excerpt in which the respondent discusses the need for nanoscale research in society and his perspective as a scientist and "a secular humanist" (0064) on the value of that research to society broadly:

Excerpt 8

1	*(0064)*	[I]n terms of social and society, I'm a humanist, a secular humanist (hh) so
2		I think it's important to do what you can in general to make this a better
3		place to live. For all sorts of people to live and that includes what do we
4		do.... Cellphones and all kinds of computational possibilities and things
5		that can be done in industry to make things better, easier and faster, all
6		those things are going to help. At some level. But the major problem we're
7		going to be—and that's from my point of view, being a secular humanist
8		and not someone who's interested in technology per se. For its own sake, I
9		mean. I do drive cars, let's say that.... I'm just saying that there is a major

10	problem that's gonna get worse and worse... there's a problem as how
11	to make life as good as possible at a biological level for everybody in the
12	world. That means that diseases that are currently prevalent have to be
13	dealt with, and the question is, how do you deal with that?
14	....And <u>that</u> I would say when I think about society, that's what I
15	think about. I think it would be nice to have smaller, more fuel-efficient
16	cars of course (hh) but if there were no cars at all there would be still be
17	people and people would still have these problems. And they're going to
18	get sick and die and that's a major problem....And what could nanoscale
19	work do to help in that regard. That's a pretty good question.

The respondent quoted earlier clearly understood the role of the scientist in society as "to do what you can in general to make this a better place to live" (lines 2–3). Although he concedes that better cellphones and "smaller more fuel-efficient cars" (line 15) have their place in this scenario, he makes it clear that his reference to "mak[ing] this a better place to live" indicates activities such as curing disease (lines 10–12), an area that falls within the domain of nanoscale research (lines18–19) and, by extension, nanoscale researchers. Earlier in the interview he had stated that it would most likely be "people in business" that were interested in nanotechnology, adding that he thought that was "important" and "great," but that the real social issue that nanoscale research needed to address was how to prevent people from getting sick and dying. Later in the interview he spoke at length about the growing elderly population around the world and the need for medical research to reverse aging and the diseases that accompany aging.

The data presented earlier from Excerpt 8 indicate an understanding of social relations that includes a distinct role for industry in providing better technology ("to make things better, easier and faster," line 5), and a distinct role for science (and scientists) in "making life as good as possible at a biological level for everybody in the world" (lines 11–12) by curing disease and reversing the effects of aging.

Coupled with the previous materials presented in this chapter, the aforementioned analysis provides a more complete understanding of the appropriate role for scientists vis-à-vis society according to the speech code shared by the scientists and engineers I encountered during the course of my fieldwork in this nanoscience community. Adding to the earlier premises and norms presented, I offer the following additional cultural premise regarding social relations in this speech community:

(P6) The role of the scientist vis-à-vis society is to understand and discover things that help make the world a better place for everybody to live.

Public Understanding, Media Portrayal, and Education

When I first asked respondents about SEI related to nanotechnology, some responded with a reference to K. Eric Drexler's 1985 concerns about self-replicating nanobots turned into environmental grey goo, a scenario they dismissed as "science fiction" (0056), "completely ridiculous" (0075), and one that caused one respondent to "want to distance myself from those guys as much as possible" (0062). That these references were invoked by the respondents without any prodding from me indicated that the respondents were aware of a larger discourse about SEI related to nanotechnology and understood this discourse in terms of some of the more bizarre and farfetched claims made by prominent figures in the field such as Drexler.

The respondents raised concerns that science journalists and news departments are disappearing and that the media cannot be depended upon to provide thorough and accurate coverage of science and technology (0069), that people will mistake science fiction or inaccurate news coverage for fact (0056), and that when a particular area of science loses public support it also loses funding (0070). Many of the respondents called for public education as an antidote to the aforementioned concerns, illustrating another particular understanding of social relations within this speech community. In the excerpts that follow, I address each of these concerns in more detail and suggest how the respondents understood the role of society at large (including the public, the media, and educators) vis-à-vis science.

In the first excerpt I present in this section, the respondent raised a concern about inaccurate portrayals of nanotechnology in popular media:

Excerpt 37

1   *(0056)* And I think that the only thing that I see different [with nanotech] is that
2   ...with some of the movies and books and things like that that are out
3   there, that some people view the nanotech more as science fiction.

In the aforementioned excerpt, the respondent suggests that the portrayal of nanotech in popular media has led to an inaccurate public understanding of what nanotechnology is. Implicit in this statement is that concerns about nanotechnologies in the public sphere are a result of fictional portrayals and that these concerns are based on fiction, not reality. Similarly, preliminary investigations into social and ethical understandings at Cornell University's NanoScale Facility (Paul 2005), highlighted that the nanoscale researchers interviewed for the project expressed concerns about science fiction representations of nanotech. Paul reported that researchers were worried that the representations "led to high unrealistic expectations of science and technology, and has also led to fomenting fear of the fictitious devices designed in these science thrillers" (Paul, 2005, p. 105). Indeed when I asked one respondent what came to mind when he heard the phrase "social and ethical issues related to nanotechnology and nanoscience," he answered as follows:

Excerpt 38

1   *(0062)* I guess the first thing that comes to mind is some of the fanciful things that
2   people have suggested might happen if nanotechnology went awry...
3   grey slime of nanomachines...

Other respondents made references to science fiction portrayals of nanotechnology, implying that such portrayals had led, or would lead, to unwarranted public concern about nanotechnologies as the following excerpt indicates:

Excerpt 39

1 (0056) Well, I think... some of these science fiction books and movies and things
2 like that.... people worry about mad scientists and things like that.

Not all the respondents were concerned about science fiction portrayals leading to negative public reaction. Several respondents suggested that the links between nanotechnology and science fiction appealed to the larger public. For example, one respondent suggested "people like the science fiction aspect, nanorobots and stuff" (0067). Another said that the fact that the prefix "nano" was being used by Apple to market one of their iPods was proof that the public was favorable to the idea of nanotechnology, even if they did not understand it:

Excerpt 40

1 (0070) Because of this political agenda, you may want to have it named this way,
2 because the public seems to be positive overall, although they don't know
3 why....
4 Apple would have done a market search just to check this before they
5 named [the Nano iPod] this way, what the public perception was, and
6 they figured OK, this is positive, the naming of it.

Contrary to the statements that many respondents made about the inaccurate portrayal of science in science fiction, one respondent said that science fiction scenarios often proved to be quite accurate:

Excerpt 41

1 (0063) I'm not a big fan of science fiction, but you go back into old science
2 fiction and you see these things taken in context back then, they looked
3 wild, right? And here we are.

Overall, however, I detected an undercurrent throughout my fieldwork and interviews that public misunderstanding of nanoscience and nanotechnology would lead to impeding scientific progress. For example, the respondent who pointed out that Apple's use of the prefix nano indicated favorable public opinion of nanotechnology, said that he was concerned that "who's using the word nano" is "out of control" (0070) and that an inevitable product mistake will lead to a loss of public support, loss of federal support and collapse of the field:

Excerpt 42

1 *(0070)* ...one [practitioner] is going to botch something up, one company is
2 going to come up with a product which kills people. And this product was
3 labeled nano. It could be an old crappy product, just re-labeled, whatever,
4 doesn't matter, right. That brings the public perception and the public
5 opinions and then follows the politicians and this will all sink it into the
6 hole, right, and then some other field may take over. I think one has
7 seen this in many things...

Indeed, at the time the aforementioned interview was conducted, the first nano-product recall had recently occurred, a German bathroom powder named "Magic Nano," the use of which caused respiratory problems in seventy-seven people (this figure was later updated to include more than 100 people) and prompted the ETC Group, a human rights advocacy organization specializing in the impact of new technologies on poor populations, to issue a press release "renew[ing] its 2003 call for a global moratorium on nanotech lab research and a recall of consumer products containing engineered nanoparticles" (ETC Group, 2006). Although it was later determined that the product had nothing to do with nanotechnology, aside from the label (Talbot, 2006), the reactions the case prompted appear to confirm the fears expressed by the scientist quoted in Excerpt 15. Additionally, Rejeski (2008) echoed these concerns, citing the German bathroom powder scare: "it would be entirely possible for serious backlash against nano to occur here if some nano product (real or not) is linked to injury to consumers."

Thus, one of the major ethical issues associated with nanotech, from the perspective of the speech community of nanotech researchers, was that the inaccurate use of the label would lead to consumer injury and subsequent loss of public support and federal funding. Again, this issue was characterized by some of the respondents as insufficient public education. For example, one respondent identified a "major problem" of ethics as lack of public understanding of science:

Excerpt 43

1 (0064) Well, the ethical issues have to with how you would—I mean the idea of
2 understanding something better... I think your major problem is that
3 there's a big—well, there—we have so many problems, particularly in this
4 country, cuz there's a real disconnect, I think, between science and
5 society.... There's a major disconnect. Where would you learn about
6 science. Hmm? On TV?
7 ....I think the media...are quite distant....the problem when you're in
8 the media is that you have to come up with something that catches
9 people's attention.... and so the kinds of problems that we're talking
10 about right now are not things that you can deal with on a...news hour,
11 let's say....
12 (DB) What would improve that?
13 (0064) Well, there's got to be some kind of connection...a better connection
14 between science and society.... And you really have to turn—I think
15 you have to turn education around.... give...a picture of reality that's
16 every bit as real as a Hollywood movie set.

Similarly, another respondent cited public education as a "social and ethical issue" she believes is relevant to nanotechnology:

Excerpt 44

1	*(0069)*	I think that [social and ethical] concerns can be framed in a lot of very
2		meaningful ways... educating the general public in a meaningful way
3		about science and technology as it's going to be carried out in the 21st
4		century. I think it's a <u>huge</u> deal. We've <u>clearly</u> got a <u>massive</u> problem in
5		this country in terms of educating the general public in science and
6		technology.

These respondents characterized "the ethical issues" (0064) and "[social and ethical] concerns" (0069) of nanotechnology as "the idea of understanding something better" (0064) and "educating the general public in a meaningful way about science and technology" (0069) and used the following expressions that indicated dissatisfaction with the state of this understanding and education:

[a] major problem (0064)
    a real disconnect... between science and society (0064)
    a major disconnect (0064)
    a <u>huge</u> deal (0069)
    a <u>massive</u> problem (0069)

The following references indicate the source of the aforementioned problems and suggest clues as to what social relations are considered appropriate within this speech code:

the media... are quite distant (0064)
    there's got to be... a better connection between science and society (0064)
    you really have to... turn education around (0064)
    [students need] a picture of reality that's every bit as real as a Hollywood movie set (0064)

> We've *clearly* got a *massive* problem in this country in terms of educating the general public in science and technology (0069).

For these respondents, one of the main SEI they identified related to nanotechnology was the unsatisfactory relationship between science and the public. As the aforementioned references illustrate, they attributed this relationship to the failure of the media and the US educational system. These references indicate that the media and the US educational system are not fulfilling their appropriate role vis-à-vis science, that is, communicating science to the public. I will return to a discussion of this element of the speech code in the next chapter.

Additionally, some respondents said that lack of public support for science resulted in lack of funding for their research. Considering the data presented earlier regarding the role of the scientist in society, it would seem reasonable to conclude that for this speech community, lack of public support and lack of funding lead to thwarted progress in improving the human condition. Specifically, two respondents cited lack of public (and federal) support for stem cell research and GMOs as obstacles to their research. For example, one of these respondents described molecular biology as "over-regulated" saying "it's virtually impossible to get under reasonable conditions, reasonable experiments, to get approval" to conduct research with GMOs (0070). He added that the case of GMOs was "a good example of where, they [regulatory agencies], because of the large public concern, they overshot..." (0070). Thus, public concern led to over-regulation, and impeded scientific progress.

Concerns regarding lack of public understanding impeding scientific progress are perhaps understandable, particularly among scientists at the UW where on May 21, 2001, members of the ecoterrorist group the Earth Liberation Front firebombed the UW's Center for Urban Horticulture, presumably because of their opposition to what they believed to be genetically-modified poplar trees that were being grown in the center (Bernton and Clarridge, 2006). The Seattle Times quotes UW professor Toby Bradshaw as saying "The truth is that we're dealing with a bunch of misinformed...folks who don't understand the research that was being carried out" (Bernton and Clarridge, 2006). The conclusion Bradshaw draws in this larger popular discourse regarding the relationship of public understanding to scientific progress is illustrative in understanding the materials I present here. Situating the present discourse in the larger discourse represented by the popular press story, I propose the following cultural premise for the speech community I investigated in this study:

> (P7) Public concern about science is often based on misunderstanding science and this misunderstanding leads to impeding scientific and technological progress.

Additionally, I submit the following cultural premise regarding the appropriate ways of recognizing the value of scientific research according to this speech code:

> (P8) Science should be supported through public education, accurate media portrayal, and federal funding.

Conclusions

The materials I collected illustrated the following categories of statements related to social relations: (1) the importance of public interest and support, (2) the availability of funding, (3) the perceived value of basic research, and (4) accurate portrayals of nanotechnology in popular and news media. These four areas were interconnected for this community, for example, public interest and support was closely linked to the availability of funding by those to whom I spoke. Similarly, concern about accurate portrayals of nanotechnology in popular and news media was related to concern about public awareness and acceptance, which, in turn, was linked to funding. An over-riding theme in these areas was that of the perceived value, and understanding, of basic scientific research.

CHAPTER 4

Strategic Conduct in a Code of Science

Abstract In addition to beliefs about human nature (psychology) and social relations (sociology), a speech code indicates distinctive beliefs about strategic conduct (rhetoric). This chapter concludes the presentation of three elements of the nanoscience speech code discovered with a discussion of the speech code within the nano community regarding the appropriate role of strategic conduct in the research and development of emerging science and technologies.

Keywords Ethnography of communication · Nanotechnology · Social and ethical issues · Speech codes theory

Overview

In the previous two chapters, I have presented data from my interviews and fieldwork, as well as from secondary sources, in order to consider aspects of the speech code in use in the nanoscience community regarding understandings of human nature (psychology) and social relations (sociology). In this chapter I conclude this presentation of elements of the nanoscience speech code with a discussion of the speech code within the nano community regarding the appropriate role of strategic conduct (or rhetoric) in the research and development of emerging science and technologies.

In addition to beliefs about human nature and social relations, a speech code indicates distinctive beliefs about strategic conduct

(Philipsen et al., 2005, p. 61). Here, strategic conduct refers to "the appropriate and efficacious symbolic resources available to interlocutors for constituting themselves as persons in social relationships" (Philipsen, 1992, p. 127). In what follows, I will present evidence for a specific understanding among users of the science code I identified for appropriate uses of strategic conduct concerning science.

The chapter is organized in the following way. First, I will present a series of data sets created from a larger body of materials that includes the interviews I completed with NSE researchers and my fieldwork activities. Then I will orient to key features of that data that pertain to beliefs about the appropriate (and inappropriate) use of strategic conduct and suggest how these features further inform the code of science I have explicated in the previous chapters. In closing, I will consider an excerpt that employs elements of all three of the related areas a speech code embodies: psychology, social relations, and rhetoric. I begin with a discussion of appropriate communication about science.

Appropriate Communication about Science

Although scientists clearly *do* consider broad SEI, as materials presented in Chapter 3 illustrate, it is not part of the dominant code of communicative conduct I documented in this speech community to bring attention to these issues by talking about them. As my classmates told me during a nanotechnology course, why would a scientist want to bring attention to what they called "negative" (0081) aspects of their work (e.g., broad social and ethical implications)? It is important to highlight the positive aspects in order to get published, secure funding, and gain public support, they told me. Similarly, writing about these issues is not appropriate communicative conduct because, as one scientist explained to me off record, scientists would not get published if they pointed out negative aspects of their work. Drawing on findings presented in Chapter 4 (pp. 106–108) related to what I called a more human nature of a scientist present in the speech I documented, specifically that scientists were motivated by recognition, funding, and fame in addition to the altruistic motivations I documented in the speech, I formulated the following rule pertaining to appropriate communicative conduct for a scientist that helps explain why those to whom I spoke were not disposed to discuss social and ethical implications of their work:

(R2) A scientist should not emphasize "negative" aspects of his/her work

Additional examples of inappropriate use of strategic conduct regarding science included science fiction writers, as well as popular nonfiction writers on nanotech such as the aforementioned K. Eric Drexler, Bill Joy, and Ray Kurzweil. In what I understood as a reference to these writers, one scientist said that a lack of understanding exists about nanotechnology because some "people on the fringes... oversold it" (0076). In the excerpt that follows, I had just told the respondent that I had read that nanotechnology was going to be the next Industrial Revolution. He responded as follows:

Excerpt 1

1 *(0076)* It's rhetoric. I don't think people actually <u>doing</u> the science had any
2 unrealistic expectations. It was people on the fringes, and journalists,
3 and some individuals who, yeah, who oversold it.

Thus, according to this respondent, "people on the fringes, and journalists, and some individuals" not "people actually *doing* the science," that is, scientists, made this claim. It is actually key figures in nanotechnology such as Mihail Roco, the National Science Foundation's senior advisor on nanotechnology, and Neal Lane, the Clinton administration's science and technology adviser, who have repeatedly referred to nanotechnology in terms that evoke the next Industrial Revolution (see Sparrow, 2007, esp. p. 60; Berube, 2006), rather than those figures on the fringes such as Joy, Drexler, or Kurzweil. Significantly, Roco and his colleagues garner respect within the scientific community. I base this assertion in part on the fact that several articles authored by Roco and his colleagues were included in the assigned reading packet I received in a science and engineering class I took at the UW. Additionally, as the architect of the NNI, Roco is indisputably a valuable advocate to scientists for allocating federal funding to nanotechnology research and development. Without exception, those I spoke to involved in nanotech research disagreed with the characterization of nanotechnology as the next Industrial Revolution. Despite the fact that scientists such as Roco are actively promoting such characterizations, no one I spoke to for this project criticized Roco or other prominent mainstream scientists, but rather criticized K. Eric Drexler for introducing the term "grey goo" in 1986 to the world and Bill Joy for publicizing Drexler's fears in a cover story for *Wired* magazine in 2000.

Thus, there appeared to be a subtle distinction between appropriate and inappropriate use of rhetoric among those I talked to for this project and I found it difficult at first to distinguish between the two. As I analyzed my materials, I began to suspect that the distinction was closely tied to the results within the scientific community which the rhetoric produces. For example, the rhetoric of figures like Drexler, Joy, and Kurzweil, is rejected by those in the scientific community perhaps because it resulted in widespread concern about the social and ethical implications of nanotechnology, as the later analysis will indicate. Additionally, inappropriate rhetoric can be viewed as that which brings on the "ethics police" (0077), a term used by one of the ethicists I interviewed for this project. In Chapter 5 (pp. 162–164), I presented respondents' comments and concerns about what one called "science fiction" (0056) portrayals of nanotechnology with implicit and explicit references to figures such as Michael Crichton, K. Eric Drexler, and Ray Kurzweil, only one of whom (Crichton) actually writes fiction. I revisit these comments in the following two excerpts to illustrate the type of communication considered inappropriate by the respondents. In the excerpt that follows, the respondent had explicitly referred to Crichton's book *Prey* immediately preceding these comments:

Excerpt 2

1 *(0056)* I think that's the challenge, the nanotech challenge in the interface that
2 way, is getting the the public to understand, you know, what's really truth,
3 what's scientifically sound, what's just good entertainment, and what's
4 just completely bogus kind of science fiction deal that that masquerades
5 under science fiction.

Of particular salience to me as I analyzed the aforementioned excerpt, was the respondent's reference to Crichton's work as "just completely bogus kind of science fiction that . . . masquerades under science fiction" (lines 4–5) and I wondered why the respondent differentiated between "scientifically sound" science fiction, science fiction that was "good entertainment," and "completely bogus" science fiction. I will revisit this matter shortly, but first present an excerpt in which the respondent refers to both Drexler and

Kurzweil as "nuts." I had told the respondent about seeing futurist and nanotech enthusiast Ray Kurzweil talk about nanotechnology leading to human immortality. The respondent answered as follows:

Excerpt 3

1 (0062) Yeah, yeah, so immortality and like um repairing frozen human brains so
2 that they can think again a hundred years about? That's the stuff I was
3 talking about where the nanotechnology books? Especially the early ones?
4 Those guys are nuts. I mean—
5 (DB) Yeah, like Drexler and those guys?
6 (0062) Oh god, yeah, yeah, I don't—you know—that stuff I want to distance
7 myself from very much because that's just complete—yeah.

Turning again to the SPEAKING framework, consider the following terms referring to Participants:

>people on the fringes
>journalists
>some individuals
>[not] people actually <u>doing</u> the science
>science fiction [and its authors]
>Those guys [i.e., Drexler, Kurzweil, etc.]

The following terms referring to Acts from the aforementioned excerpts indicate clearly what these respondents think about the communicative conduct pertaining to nanotechnology made by the aforementioned Participants:

>[employing] rhetoric
>[having] unrealistic expectations
>oversold [nanotechnology]
>masquerades under science fiction

Combining terms for Participants with the terms for Acts from the aforementioned excerpts helps explain why the communicative conduct was rejected by the respondents. This motley group of "journalists," nonscientists, fringe individuals, and science fiction writers employed "rhetoric" that created "unrealistic expectations" that "oversold" nanotechnology and was not "scientifically sound." Respondent 0056's implication that Crichton's *Prey* was neither "scientifically sound" nor "good entertainment" but rather "masquerade[ing] under science fiction" suggests that, for users of this speech code, even science fiction writers are expected to employ rhetoric that serves to advance the goals of science. As Sparrow (2007) has noted, proponents of nanotechnology must maintain a careful balance between exciting the interest of funders and the public in this "revolutionary" technology and assuring them that there is nothing to fear (p. 61).

Sparrow examined the rhetoric of a disparate group of prominent individuals and groups in the nano community that he referred to as nanotechnology enthusiasts and found that "two pairs of contradictory narratives" (p. 57) were used in the speech of these individuals and groups. The first narrative involved both "revolutionary *and* familiar" (p. 57) claims about nanotechnology. Sparrow argued that the former is present in rhetoric from seemingly disparate figures such as Drexler and Roco and focused in particular on the claims from each that nanotechnology will "change the world" (p. 59). The latter narrative, that of the familiar, is used, according to Sparrow, when concerns about these revolutionary technologies are raised:

> This narrative emphasizes the continuity of the new nanotechnologies with what has gone before. There is on this account nothing to be afraid of because the new technologies are just more of the same. What was touted as revolutionary turns out to be familiar. (61)

References to nanotechnology as familiar were prevalent throughout my fieldwork with eight out of twenty respondents explicitly referring to nanotechnology in familiar terms. The following quotes referencing the familiar nature of nanotechnology were obtained from my corpus of interviews with scientists and engineers:

> …you know, you go to a petroleum refinery, **they've been doing nanotechnology for decades**, right, and refining the gasoline that we put in cars to drive around and nobody's worried about that (0056).

> …all the gold collides that were used in vases and things like this, you know, **from antiquity, that's all nanotechnology, nanoscience**. People don't worry about that (0056).

...the petroleum industry, **they've been using nanoparticles for decades** (0056)

...**[nano] technology has been here for thousand of years** and impact a lot of what used to [be] called metacolloidal products are really nanotechnology products and they have been used in paints and pigments and sunscreen, you name it (0058).

Well, let's see, I've been doing science professionally now for 37 years and I have never done anything else but work in nanotechnology cause every chemist in the entire world, that's all they do.

Chemistry is all at the nanoscale, it's always been (0060).

I would say **I've been working in the quote unquote nanotechnology field then for the whole 25 years** [of career as a scientist] but I probably have only called it that for ten or so? (0061)

...**there's no big difference in my mind between nanotechnology and chemistry** all along, you know (0066).

...from the physics perspective **there's not much new in a way in nano** (0069)

nanotechnology, I think, got hyped up maybe in the last probably 6, 7, 8 years, but I think **I started working on it maybe 15 years ago** (0070).

Nano is not something new (0075).

The bolded statements in the aforementioned quotes indicate clearly that for the eight respondents who made the aforementioned comments, nanotechnology is familiar, not revolutionary. Yet, the revolutionary rhetoric surrounding nanotechnology continues, much of it from the scientific community, as Sparrow indicates. Why would the same speech community use the two contrasting narratives? Sparrow suggests that the narrative of the revolutionary is necessary (or perceived as such) to garner public enthusiasm for nanotechnology, and by extension, I would suggest, to secure funding. This might explain why none of the scientists or engineers I interviewed or encountered during my fieldwork criticized figures like Roco who employ revolutionary rhetoric but secure funding for the field, thus leading me to conclude that acceptable rhetoric related to nanotechnology was that which secured funding and unacceptable rhetoric was that which threatened funding.

The second pair of contradictory narratives that Sparrow analyzed was that "the development of nanotechnology is inevitable" and "that it is precarious" (p. 62). Again, Sparrow showed how the rhetoric of inevitability is present in both the writings of figures like Drexler and Kurzweil and in more "'mainstream' literature" (p. 59) from organizations such as the National Science and Technology Council. Sparrow concluded that technological determinism, "the inevitability of technological 'progress'" (p. 63), is a widely held view in the nano community. He suggests that the contrasting narrative of precariousness, "that unless we take action now, we will miss out on the (potential) benefits of nanotechnologies" (p. 64), is both "an important strategy in winning more government funding for research" and for creating "a legal—but also a moral—context which enables research (and product development) to proceed" (p. 64). Sparrow explains:

> The law may impact directly on research and/or manufacturing by rendering it, or procedures involved in it, illegal. Thus, for instance, some stem-cell researchers ordinarily based in the US have had to work overseas to pursue lines of research prohibited by law in the United States. (p. 64)

Sparrow's analysis of the strategic goal of the rhetoric of precariousness evokes the comments made by one of my respondents concerning restrictions on stem-cell research that I presented in the previous chapter (p. 169). Recent research conducted by the University of Wisconsin on public attitudes toward nanotechnology indicates that nearly 30 percent of Americans surveyed found the technology "morally unacceptable" (Worthen, 2008). The University of Wisconsin researchers concluded that "fitting the moral implications of nano breakthroughs into their existing belief or value systems is much more important for some groups in society at the moment than understanding the science behind it" (Scheufele, 2008). This conclusion is consistent with Sparrow's suggestion that "significant changes in public attitudes may be necessary before research and development of a technology can proceed" (p. 65). Sparrow added that:

> Sometimes the plea for change is directed towards public ears; unless "we" overcome our fears about nanotechnology, for instance, we will not reap its benefits. More often the plea is made for the government, or scientists themselves, or science communicators, to "educate" the public. (65)

Sparrow concludes that such pleas are indications that "the development of these technologies is not inevitable at all" (p. 65). I include this passage here, however, not to raise the issue of technological determinism, but in order to substantiate my own findings regarding the appropriate role for strategic conduct in educating the public about the benefits of nanotechnology. I am suggesting that within the code of science used among those Sparrow would call nanotechnology enthusiasts, and the larger community of science and technology, the only appropriate role for strategic conduct is that which promotes science. By promotion of science, the code of science means obtaining public support, creating the legal and moral context Sparrow describes, and securing funding for conducting scientific research.

For example, one of the scientists I interviewed described a practice where scientists conducting basic research write proposals in which they agree to pursue specific applications that they have no intention of pursuing in order to receive funding that favors applied research. In the excerpt that follows, he explains the changing climate of scientific research funding that necessitated this practice:

Excerpt 4

1 *(0070)* I think this whole trend internationally, I would say, started in the U.S. and
2 then what I observed in Europe was that there was lot more reluctance to–
3 Also because the research funding mechanism is different. Fortunately...
4 there is still a little bit-not much left-of an old-fashioned, I would say,
5 European tradition...we fund research (3.0) for (2.0) culture reasons. And
6 this I think is extremely...look at our Congress and our Senate, they
7 wouldn't do it, you need to tell them what is...the pay off. And so, this is
8 why I think there is a lot more tendency to invent new things that you can
9 communicate to more pragmatic initiatives on existing problem[s]....
10 [T]hen this changed in [Europe]...away from sort of fundamental

11	research... tight budget in the government also led to the scientists
12	becoming more creative and uh oh, then we're not doing fundamental
13	research, we need to do applied research.

The aforementioned excerpt echoes several points discussed in previous chapters, particularly the tension between basic and applied research and the perceived social value of each. For example, he refers to a quickly disappearing "old-fashioned... European tradition" of funding research "for culture reasons." He describes what he calls an international "trend" that "started in the U.S." that required federally funded research to have a "pay off," or application. For example, in the United States, he says, "you need to tell [Congress and Senate] what is... the pay off." When budgets tightened in European federal funding, he said, scientists began to frame their research in more applied terms. This required scientists to "invent new things that you can communicate to more pragmatic initiatives on existing problem[s]" and to become "more creative" about describing their research. Although difficult to convey in a transcription, the phrase "uh oh, then we're not doing fundamental research, we need to do applied research" was stated in a way that evoked surprise (uh oh) and a linguistic shift when talking about the same research. This is supported by the subsequent comments this respondent made:

Excerpt 5

1	*(0070)*	It was actually the first thing, that the rhetoric changed in that sense, you
2		do applied research, you (don't) do fundamental research, and more
3		leading to some innovative technology which benefits society rather than
4		we want to know *whyy doo* electrons do this? Just because they're there
5		we want to know... Worked in the past, it doesn't work so well anymore
6		(hh).
7	*(DB)*	So, overall, you're saying there's a trend towards applied, applied science,
8		applied research.
9	*(0070)*	In the way one communicates.

He notes that it was "the rhetoric" that changed first. For example, he says although it had "[w]orked in the past" for scientists to study something "[j]ust because...we want to know," that this justification "doesn't work so well anymore." Instead, as a scientist seeking federal funding, "you do applied research, you (don't) do fundamental research, and more leading to some innovative technology which benefits society."

According to respondent 0070 quoted earlier, although scientists write these proposals emphasizing basic science, this emphasis is limited to "the way one communicates," adding that "we get this proposal approved and then we keep doing what we wanted to do," i.e. basic research (0070). He explained further:

Excerpt 6

1 *(0070)* If you're clever, you know how to phrase your proposal to get the funding
2   that's available. And then the hope is that you can do what you really
3   want to do.

The respondent told me a related anecdote about a Nobel laureate who advised young scientists to say what they had to say in order to continue doing their research:

Excerpt 7

1 *(0070)* [He said] apply for the applied research and once you get the money you
2   can still do fundamental research. So you don't have to-what he's saying
3   is just be political, don't ride your principles.

Similarly, in a separate interview, another respondent commented on the strategic use of the term "interdisciplinary research":

Excerpt 8

1 *(DB)*   ...so [you're saying] interdisciplinary research isn't new, it's just the term
2   is new.
3 *(0076)* Well, you're <u>talking</u> about it and putting a big <u>emphasis</u> on it, and...

4	telling people they <u>have</u> to be interdisciplinary, <u>that's</u> new. In the
5	olden days, people did it when they needed it (hh). Nowadays, you have to
6	do it to get funded, so (hh).... You have to talk about it a lot more.

These excerpts indicate an ethic regarding communication that is widespread among basic scientists: that this use of strategic conduct was understood as acceptable by members of this speech community because it served to advance the goals of science. However, the statement "don't ride your principles" (Excerpt 7, line 3) as well as the respondent's use of the term "have to" three times in Excerpt 8 when describing the recent practice of referring to one's scientific research as interdisciplinary, suggests some tension within this community regarding this use of strategic conduct. This tension was most clearly evident in the way the respondents I interviewed talked about the term "nanotechnology."

Many of the respondents noted that they called their work by other names (these included completely unrelated areas, for example, "atomic force microscopy" and "chemistry") prior to the introduction of the term nanotechnology and although the name had changed, their work had not. One respondent said that he felt like he had been "a little unempowered" by the fact that "someone invented a name for everything I do", adding "but in any event, it's just a name" (0060). Another respondent said that the term was "used as a big brush to kind of cover everything" (0056).

Overall, as the following data set illustrates, there was a shared, expressed "disdain" (0072) among members of this community for the word nanotechnology, although they all used it to describe their work. For example, respondents referred to the term in the following ways:

> the quote unquote nanotechnology field (0061)
> a new spin on words [that] people have taken... to mean everything and anything that they want it to mean (0060)
> largely politically driven (0070)
> a buzzword (0072), (0076)
> jargon (0076)
> fancy new words that can appeal to the public (0076)
> so called nanoscience (0076)
> a non-useful term (0074)
> a little bit like Swiss made... just a stamp (0062)

The use of words like "politically driven," "jargon," "a buzzword," and "spin" indicate why the respondents expressed distaste for the term nanotechnology. These words are suggestive of the inappropriate strategic conduct referred to earlier, what one respondent called "rhetoric" and overselling (0076). Thus, using a term like nanotechnology is, for some, in conflict with one of the rules pertaining to strategic conduct held by this speech community:

(R3) A scientist should not oversell his or her work by making unrealistic claims.

Considering the tension in this speech community around the term nanotechnology helps illustrate the overlap between understandings related to human nature, social relations, and strategic conduct. In addition to recognizing the term as a violation of a rule about appropriate communicative conduct as suggested earlier, as discussed in Chapter 3, the term violated a sense of identity held by some respondents. For example, those respondents who identified themselves as basic scientists did not like the implication of applied science that the term nanotechnology held. As one respondent said, "most people doing [basic science] don't see themselves as technologists, so we have an instinctive discomfort with the word" (0076). Thus, use of the term nanotechnology comes into tension with beliefs and values pertaining to a scientist's identity, mainly that basic science does not involve developing products for application. Finally, use of the term nanotechnology threatens appropriate social relations within this speech community by implying that the role of the scientist vis-à-vis society is to make technology rather than to "just understand things" (0076).

Not all the respondents shared the aforementioned respondents' discomfort with the term nanotechnology, however. One scientist said that although some scientists turned their noses up at the strategic use of the term, the practice of strategic naming was standard in science:

Excerpt 9

1 *(0072)* I think people's disdain for the word nanotechnology comes from the
2 fact that it was basically coined as a buzzword to get funding. But this is

3	true for most things in science, you know, is that you coin something to
4	bring a field together...

The aforementioned response, paired with the earlier excerpts, is illustrative of the belief within this community that strategic conduct that elicits funding for science is appropriate:

(P7) Strategic conduct that promotes scientific research is appropriate scientific conduct.

Cultural rules refer to how things *should* be rather than how they actually are where premises refer to the beliefs and values that a speech community holds about certain things (Philipsen, personal communication, 2008). Thus, engaging in communicative conduct that violates a cultural norm by referring to one's scientific research using a term that is considered by the community to be "spin" (0060) is an acceptable violation because it satisfies a cultural premise held by the same community that promoting scientific research is appropriate communicative conduct. In this way, I suggest, the tension evident in this community about the use of the term nanotechnology is resolved, albeit uneasily, as indicated by the respondents' reactions to the term.

An additional example of using a premise pertaining to appropriate communicative conduct to accommodate a violation of a rule pertaining to appropriate communicative conduct is illustrated in the reactions some of the respondents had to Drexler's 1986 book "Engines of Creation." This text, largely acknowledged as the first serious work on the potential of nanotechnology as well as the text that generated the study of related SEI, was consistently condemned by members of this speech community in terms such as "completely ridiculous" (0075), as illustrated in the previous chapter (p. 162). However, some of the respondents noted the usefulness of texts such as Drexler's in generating interest and funding for nanotechnology. One, for example, credited Drexler's work as "part of what motivated... the national push for nanotechnology" (0062). Similarly, another scientist said that the fantastic claims of nanotechnology are "what the congress people and the public want to hear" (0076). Again, these data suggest that members of this speech community acknowledge that communicative conduct they otherwise would characterize as inappropriate can nonetheless serve scientific interests.

As by now is likely evident, the common way of speaking related to appropriate communicative conduct within this speech community was that of generating funding for scientific research. Illustrating this way of speaking further is the next excerpt, in which the respondent describes "being able to communicate" as "a critical aspect of being a scientist":

Excerpt 10

1 *(0069)* I mean, it [interdisciplinary work] can be an issue because people don't
2   communicate well or a field is full of jargon and it can affect their ability
3   to get funding and all kinds of things. I mean, this is a critical aspect of
4   being a scientist is being able to communicate, sell, market your results.

Notice the use of "sell" and "market" accompanying "being able to communicate" (line 4). This implies a one-way communication model in which the goal of scientific communication is to "sell" science to the public.

Indeed, as Borchelt and Hudson (2008) point out, standard models of science communication have focused on educating the public about science and technology and the benefits of science and technology, with a one-way flow of information. Although experts have called for public engagement models that emphasize two-way communication between the public and scientists (see e.g., Leshner, 2006), and public engagement activities are taking place in growing numbers (as described in Chapter 1, pp. 14–15), critics charge that these efforts remain unidirectional (e.g., Borchelt and Hudson, 2008). This charge is based in part on the absence of a formal feedback mechanism incorporated within these efforts to relay the public perspectives back to decision-makers, as well as the lack of evaluative research on public engagement (Borchelt and Hudson, 2008).

In a 2008 article in *Science Progress*, Borchelt and Hudson noted that:

> [A]n erosion of public trust that began as a trickle of doubt about radiation safety and pesticides has grown to program-threatening uprisings against emerging new technologies, from genetically altered "Frankenfoods" to concern over "grey goo" in nanotechnology.

They criticized "the reliance that the science and technology community places in various 'deficit models' of interaction with the public":

> The basic assumption behind these models is that there is a linear progression from public education to public understanding to public support, and that this progression—if followed—inevitably cultivates a public wildly enthusiastic about research. But this model of scientific engagement with the public obviously isn't working.

Tracing the history of efforts to educate the public about science, Borchelt and Hudson describe a move from an emphasis on public education and an information deficit model (e.g., "if only lay people knew what scientists did... they too would support the agendas of the scientific establishment") to an emphasis on public understanding and what they call "an 'attitudinal deficit'" model (e.g., "to know us is to love us"). They argue that both models are "built on one-way flow of information from the expert to the public with very little information flowing back the other way" and that while these models have been in use since the mid-twentieth century, "[n]either public support for research nor scientific literacy increased significantly in all that time."

The latest recommendation for communication between scientists and the public, according to Borchelt and Hudson, is what is being called public engagement efforts (see, e.g., Macnaghten, Kearnes, and Wynne, 2005). A key, and distinguishing feature, of these efforts is "two-way, symmetrical communication" that results in "meaningful incorporation of public input into [the science policy] process" (Borchelt and Hudson, 2008). Although Borchelt and Hudson suggest that public engagement models are promising in theory, they criticize current efforts as continuing a "one-way, expert-to-layperson information delivery, albeit in different settings like cafes scientifique, public meetings, and town halls." Part of this problem, they suggest, is a change in naming without a change in practice. In particular, they argue for the incorporation of "institutionalized mechanisms for reflecting the public's input in deliberation or policy construction." They conclude that "the end game of public engagement should be empowerment":

> [C]reating a real and meaningful mechanism for public input to be heard far enough upstream in science and technology policy making and program development to influence decisions. It is not about making a decision

among a scientific elite, and then staging public events to move the public toward agreeing with that desired outcome. It is about empowering lay citizens to learn all they want about pending program or policy issues (not what scientists believe they *need* to know to weigh in), and then giving them access to deliberative processes where that knowledge can be questioned, applied, and incorporated with knowledge or questions gleaned from outside the scientific process.

In 2006, Alan Leshner, chief executive officer of the American Association for the Advancement of Science and executive publisher of *Science*, had published a commentary in *The Chronicle Review* in which he advocated a public engagement model between scientists and the public that emphasized listening and "genuine dialogue." He noted that "a respectful dialogue with the public is much more effective in finding common ground than a more traditional, instructional monologue."

Leshner (2006), Borchelt and Hudson (2008), and Macnaghten, Kearnes, and Wynne (2005) note that public engagement efforts are a departure from past efforts not only in the move from a one-way to a two-way information flow but also in the expectation that the goal of the efforts is not to convince the public to share the views of the science community but to understand the public's views and incorporate them into science policy. I include the discussion on public communication about science here in order to illustrate how ideas about strategic conduct, specifically communicating with the public, are positioned with the larger discourse taking place about the value of, and the method of, communicating scientific information to the public. While replacing a one-way information model with a two-way model and changing expectations regarding the goal of the model of scientific communication are not simple tasks, these challenges are further intensified by what appears to be a deeply embedded view within the scientific community regarding the appropriate strategic conduct of scientists as illustrated by usages of terms for talk in the interviews I conducted with scientists and engineers. It is to that analysis that I now turn.

Terms for Talk

One of the ways that I looked for rules pertaining to strategic conduct within the science speech code I documented was to analyze the interview transcripts to determine how the respondents talked about talk. I did this by identifying the instances in which the respondents used the following terms:

Communicate
Explain
Speak
Talk
Say
Tell

I also looked for additional terms for talk not listed earlier and although I found several, I discarded them as anomalous or unproblematic for the purposes of the present analysis.[1] Additionally, from these instances listed earlier, I discarded those that I judged unconsequential (e.g., see the discussion of "say" later). The final number of terms for talk that I analyzed in depth was 42. I looked at each usage for context and co-occurring terms that would provide insight into what the speaker meant by the term he or she used for talk. In what follows, I present that analysis and findings. I begin with an analysis of the term "communicate" which appeared nine times in the interviews and the term "explain" which appeared three times and with the same co-occurring terms as did "communicate."

Instances of "Communicate" and "Explain"

In the following excerpts, I have bolded instances of the terms "communicate" and "explain" as well as co-occurring terms such as "language."

Excerpt 11

1 *(0056)* I think probably one of the things the field [i.e., nano] needs to do is get
2 good, well-accepted **definitions** of all these things [i.e., terms related to
3 nano] so we (hh) can **communicate** a little better on 'em.

Excerpt 12

1 *(0058)* ...when you cross... even between say physics and engineering, or
2 chemistry and engineering there are some challenges of just
3 **communication, language, vocabulary**...

Excerpt 13

1 *(0067)* ...every discipline has [a] slightly different **language** that they use to

	2	**explain** their phenomenon, so sometimes we do have misunderstandings,
	3	but I think the more we **communicate** with each other the more we'll
	4	learn.

Excerpt 14

1	*(0070)*	...look at our Congress and our Senate, they wouldn't do it [i.e., fund
2		basic research], you need to tell them what is... the pay off. And so, this is
3		why I think there is a lot more tendency to invent new things that you can
4		**communicate** to more pragmatic initiatives on existing problem[s]....
5	*(DB)*	So, overall, you're saying there's a trend towards applied, applied science,
6		applied research.
7	*(0070)*	In the way one **communicates**.

Excerpt 15

1	*(0069)*	I mean, it [interdisciplinary work] can be an issue because people don't
2		**communicate** well or a field is full of jargon and it can affect their ability
3		to get funding and all kinds of things. I mean, this is a critical aspect of
4		being a scientist is being able to **communicate, sell, market** your
5		results.

Excerpt 16

1	*(0072)*	[It's important to make] a point of going and trying to **communicate** to
2		these audiences [i.e., students in other science disciplines]. I don't just
3		try to **communicate** to my colleagues, I go and give a **talk** and
4		**communicate** to their students because that's when you really understand
5		what-where people are coming from...

Excerpt 17

1 *(0075)* ....it's like you get rid of the borders in Europe, right and then nobody
2 **speaks** the same **language**...it's a similar thing here. We really **speak**
3 different **languages**. We have also a little bit developed different way of
4 approach, right, I mean, physicists [are] very meticulous...while...the
5 engineer doesn't really care about all the details, he likes the product, he
6 looks at the product side and **says** well, the product matters, right. So these
7 are two different ways of approaching it....they just have to figure out
8 how to **communicate**.

Excerpt 18

1 *(0068)* ...I think it's good for them [i.e., scientists] to go through this exercise
2 [i.e., present their research to non-scientists] for a couple of
3 reasons, one is that they have to be able to **explain** their science to non-
4 scientists, so that's for **communication** purposes...

Excerpt 19

1 *(0059)* ...a lot of the time just goes into **explaining** each other's **languages**. I
2 have been having recently meetings with bunches of doctors...trying to
3 put together a joint program using these nanoparticles for various
4 interesting bio-medical issues, and it's almost like they're **speaking** a
5 certain **language** that I don't understand and I'm **speaking** a certain
6 **language** which they don't understand.

Excerpt 20

1   *(0067)* ... every discipline has slightly different **language** that they use to **explain**
2   their phenomenon, so sometimes we do have misunderstandings, but I
3   think the more we **communicate** with each other the more we'll learn, I
4   don't think it's a huge barrier.

Excerpt 21

1   *(0063)* [What is needed for improved interdisciplinary research is] education, you
2   know, you **explain** to people what you do and you hope
3   you break through the barriers...

In the aforementioned excerpts, forms of the term "communicate" co-occur with forms of the term "explain" three times. The term "communicate" co-occurs with the term "language" four times and the term "explain" co-occurs with the term "language" three times.

Additionally, one or both terms co-occurs with the following terms at least once:

definitions
vocabulary
talk (n.)
speaks
tell
sell
market
education

What can be learned from the aforementioned occurrences to aid in better understanding the respondents' perspectives on the use and value of strategic conduct? The usage of the term "communicate," closely related to the term "explain" as presented earlier, suggests culturally distinctive understandings regarding communicative conduct in this speech community. Just as Katriel and Philipsen (1981) discovered that "communication" meant, in the corpus of speech they examined, "*close, supportive,* and *flexible* speech" (p. 315, italics in original) in a mainstream American speech code they identified in the speech of some individuals and popular culture in the late 1970s and early 1980s, so "communication" in the speech community of scientists and engineers working in nanotechnology also refers to something particular, in this case, explaining their research effectively to generate the desired result, whether that desired result is satisfactory collaboration with other researchers, obtaining funding, or securing support from nonscientists.

It is also significant that the term "language" and terms referring to language such as "definitions" and "vocabulary," co-occurred most frequently with the terms "communicate" and "explain." This suggests an understanding of communication among the users of this speech code that privileges words and meanings of those words over values such as "close, supportive, and flexible" when referring to speech. I will now turn to an examination of instances of the terms "speak" and "talk" in the interview data.

Instances of "Speak" and "Talk"

Across the interview data, forms of the term "speak" occurred five times and forms of the term "talk" occurred twenty-two times, the most of any of the terms for talk used in this data. I present the two terms as a pair following the lead of Dirven, Goossens, Putseys, and Vorlat (1982, as discussed in Philipsen and Leighter, 2007) in which they analyzed uses of the verbs "speak," "talk," "say," and "tell." Dirven et al. based their analysis primarily on a corpus of material obtained from "sixty British stage plays dating from the period 1966–72," supplemented with additional materials such as "newspaper language and informants" (p. 6). Philipsen and Leighter note that the pairing of the verbs serves the following functions:

> [They] emphasize speaking or talking in its own right.... Furthermore, within the "speak"/"talk" pair, "speak" differs from "talk" in that the former is more likely to perspectivize linguistic action as a more unidirectional act from a speaker to a receptor...whereas, "talk" is more likely to perspectivize the addressee as a potential interactor.... (207–208)

Later I first present an exhaustive list of the references to forms of "speak" or "talk." I then suggest what is significant about the usage of these two terms in this data.

Excerpt 19
1 (0073) ...you have to spend a lot more time **talking** with people....
   scientists
2   from different fields **speak** different **languages** and you have to spend
3   time to get to know about that person's discipline and about how that
4   works and it's very different usually from how your stuff works, and
5   getting that **dialogue** and beginning to understand each other's work, is
6   extremely important...

Excerpt 20
1 (0075) ....it's like you get rid of the borders in Europe, right and then nobody
2   **speaks** the same **language**...it's a similar thing here. We really **speak**
3   different **languages**. We have also a little bit developed different way
4   of approach, right, I mean, physicists [are] very meticulous...while...the
5   engineer doesn't really care about all the details, he likes the product, he
6   looks at the product side and **says** well, the product matters, right. So these
7   are two different ways of approaching it....they just have to figure out
8   how to **communicate**.

Excerpt 21
1 (0059) I have been having recently meetings with bunches of doctors...trying to
2   put together a joint program using these nanoparticles for various
3   interesting bio-medical issues, and it's almost like they're **speaking** a
4   certain **language** that I don't understand and I'm **speaking** a certain
5   **language** which they don't understand.

Excerpt 22

1 *(0072)* [It's important to make] a point of going and trying to **communicate** to
2 these audiences [i.e., students in other science disciplines]. I don't just try
3 to **communicate** to my colleagues, I go and give a **talk** and **communicate**
4 to their students because that's when you really understand what-where
5 people are coming from...

Excerpt 23

1 *(0060)* ...you are **talking** to a broader range of people, so people in other
2 disciplines...often cannot understand what each other is **talking** about...

Excerpt 24

1 *(0061)* ...we need a few of you guys [i.e., social scientists] to come and give
2 invited **talks** at our [i.e., scientists'] conferences every once in a while,
3 just to spark discussion.

Excerpt 25

1 *(0068)* ...actually learning about our collaborators' research and **vocabulary** and
2 how they look at things, I think that helps a lot.... my training is in
3 physics and electrical engineering, but I spent two-three years in a
4 chemical biology department as a post-doc.. so that makes it a lot easier
5 for me to **bridge over** and **talk** to someone who's in biology.

Excerpt 26

1 *(0069)* I think we [i.e., the scientific community] have to **reach out** to a broader
2 community of potential scientists in this country and **talk** to people who

3	have different kinds of cultural experiences and can think of framing
4	questions and problems... in diverse ways.

Excerpt 27

1	*(0069)* ...it's hard **talking** the same **language**. I spend a lot of time sort of
2	repeating the same questions, repeating the same kinds of phrases to—and
3	I'm sure that the physical chemists that I **interact** [with] have to do the
4	same with me...

Excerpt 28

1	*(0070)* ...so when say we **talk** about nano, when I give [a] research presentation,
2	it very much depends whom you **talk** to. So if I address really a
3	focused scientific community, they are already at a stage where you try to
4	avoid the word nano and come back to what actually is the substance of
5	physics...

Excerpt 29

1	*(0070)* For 500 years [nano] hadn't been **talked about** and then if one **talks** to a
2	little broader audience, if one **talks** to maybe students, industry people,
3	then one uses this **terminology** a lot more because this is where you have
4	[a] more political agenda in mind when you **talk** to industry people and
5	students and, you know, you wanna give students the job prospects, nano's
6	the newest thing, to use this **vocabulary** in a little more simple-minded
7	way in getting a job-talk... I encourage my students actually to look into

8	the subject, also because if you-there are many senior scientists who when
9	I would **talk** to them about my research I better avoid the word nano.

Excerpt 30

1	*(0071)* [I] do a lot of homework before I **talk** to the other person, I think it's not
2	something that you can just put two people in the room and it will work
3	out, it's impossible, I think everybody has to the initiator (hh) to learn
4	about the other person's **terminologies**, so I spend a lot of time reading
5	before I could get to a stage where I can **talk** to the other person…

Excerpt 31

1	*(0073)* …you have to spend a lot more time **talking** with people…. scientists
2	from different fields **speak** different **languages** and you have to spend
3	time to get to know about that person's discipline and about how that
4	works and it's very different usually from how your stuff works, and
5	getting that dialogue and beginning to understand each other's work, is
6	extremely important…

Excerpt 32

1	*(0074)* People [i.e., scientists] have been working, **talking** about this [i.e.,
2	nanotech] for forever.

Excerpt 33

1	*(0075)* Now, now, up to this point, everyone was by himself. The biologist was
2	doing what he was doing and the chemist was doing what he was doing

3	and they usually did not **interact**. And they used their own **language**, so
4	they could not **talk** with each other.

Excerpt 34

1	*(0076)* Well, you're **talking** about it [i.e., interdisciplinary research] and putting a
2	big <u>emphasis</u> on it, and…telling people they <u>have</u> to be interdisciplinary,
3	<u>that's</u> new. In the olden days, people did it when they needed it (hh).
4	Nowadays, you have to do it to get funded, so, (hh)…You have to **talk**
5	about it a lot more.

Forms of the term "speak" occur exclusively with the co-occurring term "language" in the aforementioned references, providing additional support for the finding presented earlier linking language and terms for language to "communicate" and "explain." In using the term "talk," the respondents quoted earlier do indeed "perspectivize the addressee as a potential interactor" (Philipsen and Leighter, 2007, p. 208). An analysis of the addresses in each excerpt is instructive in further illustrating the function of "talk" within this speech code:

> people [i.e., scientists]
> students [in science and engineering]
> people in other disciplines
> you guys [i.e., social scientists]
> someone who's in biology
> potential scientists
> physical chemists
> a really focused scientific community
> a little broader audience
> students [in science and engineering]
> industry people
> industry people
> senior scientists
> the other person [scientist/engineer in another discipline]
> scientists from different fields
> the biologist
> the chemist

With the exceptions of "a little broader audience" (which presumably includes industry) and "industry people" (which occurs twice), the remaining fifteen addressees aforementioned refer to scientists. This suggests that for users of this speech code, scientists and engineers, the appropriate addressee with whom to interact is, primarily, other scientists and engineers or potential scientists and engineers (e.g., students). Excerpt 24 is significant in that it is the only reference to talk that involves social scientists as the initiator of the talk. Additional terms co-occurring with the occurrences of "talk" aforementioned that suggest interaction include the term "interact" which occurs twice and the related terms of "bridge over" and "reach out" which occur once each in the aforementioned references.

Instances of "Say" and "Tell"

Philipsen and Leighter (2007) analyzed eight uses of the term "tell" in the documentary film "Corporation: After Mr. Sam." They found that use of the term was consistent with extant data suggesting that the term "tell" is used strategically by a speaker primarily for information purposes and does not imply interaction or response from an addressee. Dirven et al. (1982) suggest that the term "say" "does not necessarily involve an addressee" (Philipsen and Leighter, 2007, p. 208). Additionally, Philipsen and Leighter identified the use of "say" as indicating "concession" where the use of "tell" indicated strength (p. 219). The occurrences of "say" in the transcripts served a relatively unproblematic usage (e.g., "I think this whole trend internationally, I would **say**, started in the U.S."). The one reference that deviated from this usage includes both "say" and "tell" juxtaposed to one another in a particularly illustrative excerpt referring to scientists' hypothetical response to questions about SEI related to nanotechnology:

Excerpt 35

1 *(0065)* ... I don't think you're gonna get people [i.e. scientists] **saying** that... they
2 have big concerns.... they will undoubtedly **tell** you that there should be
3 proper regulation....

In this excerpt, the respondent says in effect that scientists do not "say;" they "tell." Considering this usage in light of Philipsen and Leighter's findings, I would suggest that, for this speaker, the passive, and concessional, suggestion of "say" is inappropriate to a scientist discussing implications of his or her work. More appropriate is the "strong or assertive" (Philipsen and Leighter, 2007, p. 219) quality of "tell" that discourages feedback. A second use of "tell" further confirms the aforementioned suggestion that the appropriate linguistic action for a scientist is to "tell" rather than to "say":

Excerpt 36

1   *(0070)* . . . look at our Congress and our Senate, they wouldn't do it [i. e., fund
2   basic research], you need to **tell** them what is . . . the pay off.

Again, the use of "tell" here suggests that scientists must take an assertive linguistic stand when discussing their research. The third, and final, use of "tell" that occurred in the interview data appears in the following excerpt, which was presented earlier for the occurrence of "talk":

Excerpt 37

1   *(0076)* Well, you're talking about it [i.e., interdisciplinary research] and putting a
2   big emphasis on it, and . . . **telling** people they have to be
3   interdisciplinary, that's new. In the olden days, people did it when they
4   needed it (hh). Nowadays, you have to do it to get funded, so, (hh) . . . You
5   have to talk about it a lot more.

This usage differs from the ones aforementioned in that in this reference, the scientists are the ones being "told" something by, presumably, department heads, funders, etc. In an earlier chapter, I noted the respondent's displeasure expressed earlier in his discussion of the naming of research as "interdisciplinary." Similarly, his use of the term "telling" suggests displeasure. Again, it is a strong and assertive term that does not encourage interaction. Thus, "people" (i.e., scientists) are not invited to provide their input on the idea of interdisciplinary research; they are simply told that

that is what they "*have* to be" in order "to get funded." Additionally, the respondent's usage of "telling" suggests that scientists, who are accustomed to being the ones doing the "telling" do not enjoy being in a reversed role such as the one he describes. This is consistent with the earlier analysis on the identity of a scientist and one respondent's statement that "we [i.e., scientists] are not *easy* to control...we don't see ourselves as employees....we see ourselves as freelance..." (0075), suggesting that scientists do not like to be controlled.

Many of the scientists I interviewed and encountered during my fieldwork appeared to understand communication solely in terms of a one-way, marketing model as described earlier. This understanding extended to the way social science research on communication, such as this project, was perceived. In what follows, I present four examples related to how scientists I encountered or observed during the course of this project understood communication and communication research related to science and public communication of science.

Four Anecdotes

When I signed up for a series of student-run nanotech seminars in order to better familiarize myself with the range of research in nanotechnology being conducted by emerging scientists and observe how this group talked about their work (which is how I presented myself to the professor and the class) the professor introduced me as someone who was studying how scientists could learn to communicate their research more effectively (0085). After I gave my presentation on my research in progress, I observed that the professor, who had previously been friendly toward me, became standoffish, not responding to e-mails and ignoring me when our paths crossed. As I continued to analyze the materials I was collecting, I understood this change in behavior as the result of my violation of a rule concerning communicative conduct. In other words, I had engaged in inappropriate communicative conduct pertaining to science. Analyzing the way scientists talk about science was not perceived by this professor to be consistent with the goals of appropriate communicative conduct within this speech community, that of advancing science, and may even have been seen as hindering science.

The second example took place while I was visiting another university to interview for a faculty position. One of the university administrators, a scientist, told me that she was excited to hear about my work in communication and nanotechnology as they were launching a nanotechnology undergraduate degree program at her university and she envisioned my communication research as helping to "market" the new program to incoming students. It became clear to me during our one-hour conversation that this individual did not understand communication research in terms other than marketing.

Third, during the 4 years I spent as a research assistant on a social science team funded to address social and ethical aspects of nanotechnology as part of a larger multi-institutional effort, I observed periodic tense interactions between the project's principal investigator (a scientist) and our team. These tense interactions concerned competing perspectives of our team's role in the overall project. For example, in one series of interactions, the principal investigator suggested that our team's communication research efforts could be best utilized by creating a user satisfaction survey for the UWNanoTech User Facility (NTUF), a research facility that houses several expensive imaging and fabrication instruments for nanoscale use by both UW researchers and external researchers for a fee. The survey would be useful to the overall efforts of the project, the principal investigator explained, which was promoting the user facility to regional industry users. He also suggested that a communication research project our team had designed to assess interdisciplinary communication across the network lacked scholarly value. The only appropriate role for communication research on SEI related to nanotechnology was envisioned by this individual in terms of marketing, specifically to promote the NTUF and generate revenue.

The fourth, and final, example occurred toward the end of my fieldwork activities. Pacific Science Center, Seattle's science museum, hosted a public forum on SEI related to nanotechnology as part of a broader national effort funded by the National Science Foundation to educate the public about nanotech in the spirit of the public engagement events discussed earlier in this chapter. The forum attracted approximately fifty people and involved an introduction by a researcher from the UWNTUF, the viewing of a 20-minute fictional film produced by NSF depicting two perspectives on whether nanoparticles posed risks to humans and to the environment, and a subsequent small group discussion. Although the content of the film shown is worthy of a separate analysis detailing the distinct ways in which each of the two sides presented were framed, it is beyond the scope of this present analysis. I will note, however, that one of the attendees at my table expressed her displeasure with what she called an "obvious bias" in the film which involved minimizing concerns about the safety of nanoparticles. The CEO of Pacific Science Center was seated at my table and he challenged this statement by saying he thought the film was "balanced." He then suggested we go around the table and introduce ourselves and when I introduced myself as a communication researcher, he smiled broadly and asked me if I was a science journalist, adding that "we really need good science journalists." When I answered that I was not a science journalist, rather that I was studying the way scientists talked about their work, he stopped smiling and looked confused. After the discussion, which was focused on questions provided on a handout about who should be responsible for regulating nanotechnologies, the groups reconvened. After each group reported to the larger group what they had discussed, the facilitator invited questions. One of the attendees then asked the following question, which I have paraphrased from my notes: Now that we've spent all this time talking about our ideas, what now? Do our comments go somewhere? Who's listening to us? This comment appeared to catch the facilitator off-guard and she deflected the question to the CEO. He answered that the goal of the evening's forum was simply to generate dialogue and nothing more. Attendees were thanked for their participation and the forum was concluded.

In addition to providing an example of the sorts of public awareness events renamed public engagement events as described earlier, I include the aforementioned anecdote because it, as well as the previous three, helps to illustrate the distinctive understanding of the appropriate role for strategic conduct in the speech code of science I documented. Namely, that the only appropriate function of communication about science (and communication research about science) is to promote (or "market") science, whether that is to incoming students, funders, or the public. Moreover, this understanding does not appear to be restricted to communication research, but extended to all social science research. For example, Macnaghten, Kearnes, and Wynne (2005) pointed out language in reports issued by the USNNI that indicated "an envisaged role of the social sciences... as a social lubricant in the drive toward industrial success and commercialization" (p. 7).

In the excerpt from the interviews that follows, the respondent had been discussing with me ways to increase collaboration between scientists and social scientists in order to address social and ethical implications of emerging science and technologies. He had suggested that a "joint forum" be held in which scientists could "present their work and then open it for a critical review by the audience" (0068). I asked him if such a forum would make scientists feel defensive and he answered as follows:

Excerpt 38

5   *(0068)* No, I think it's good for them to go through this exercise for a couple of
6   reasons, one is that they have to be able to explain their science to non-
7   scientists, so that's for communication purposes, that's actually good then,
8   and, the second thing is, that skill is necessary for fundraising, so it's
9   actually good for a scientist to go through this exercise, because as they go
10 through their career and they need to raise funds, they have to go through
11 this a lot of times, so that's a good thing to do anyway, and, then, the
12 most important thing is that if there's something that they've overlooked
13 in terms of the ethical implications of their research, they may actually
14 get another opinion about it and start thinking about it.

The respondent characterizes scientists communicating with nonscientists in this way as an "exercise" (lines 1 and 5) that "is good for a scientist" (line 5, also lines 1, 3, and 7). The reason such an "exercise" is "good" is that scientists

"have to be able to explain their science to non-scientists...for communication purposes" (lines 2–3);
   "that skill is necessary for fundraising;"
and,
   "the most important thing" (line 8): "if there's something that they've overlooked in terms of the ethical implications of their research, they may actually get another opinion about it and start thinking about it" (lines 8–10).

Consistent with the materials and data presented earlier in this chapter, the aforementioned data suggest that the primary function of communicative conduct among scientists is to generate funding. The respondent's comment about "the most important thing" (line 8) notwithstanding, "get[ing] another opinion" and "thinking about" (line 10) social and "ethical implications of their research" (line 9) follows "communication purposes" (line 7) and "fundraising" (line 4). In the section that follows I will discuss this aspect of the code further.

STRATEGIC USE OF INTERVIEW BY THE RESPONDENTS

As I conducted the interviews, I became aware of the strategic use of communication by those I interviewed. Briggs (1986) notes the importance of attending to how the interview itself is perceived by the interviewee, pointing out that the goals of the interviewee are not necessarily those of the interviewer (see p. 103). By attending to the participants and ends as I analyzed the interview data, I began to notice that nearly a quarter of the participants emphasized particular areas of concern (primarily related to toxicology) of nanotechnology. At first I was impressed with the candor that I felt was being expressed to me, particularly in light of some of my other findings. However, when I was invited to participate in putting together a proposal for funding that would involve establishing a new center dedicated to the study of toxicology related to nanotech, I noticed that the names of many of those whom I had interviewed were listed. I chuckled to myself as I realized that I had been an unwitting recipient of the type of scientific communication that the participants themselves had described for me. That is, I was perceived by some of the participants as being in a role that would involve publicizing the need for certain allocation of funding. Thus, it was entirely consistent with the scientific code of communication that the interviews with me be utilized strategically as persuasive appeals for funding.

Similarly, I detected that another quarter of the respondents appeared to be motivated to convince me that there were no SEI associated with nanotechnology. I base my assumption on the fact that these respondents adamantly denied that there were any issues even though I knew at the time or later discovered that many of them were involved in research that other respondents identified as involving serious health and safety issues. In fact, I believe if I had interviewed these respondents during the time that the aforementioned grant proposal related to toxicology and nanotechnology was being developed, their answers would have been similar to the previous group of respondents. However, perceiving the interview as a possible threat, the communicative conduct that took place was consistent with a cultural rule governing appropriate communicative conduct in this speech community, that a scientist should not emphasize negative aspects of his or her work.

Regarding the remaining interviews, I detected that many of the participants did not appear to have any agenda at all in speaking to me, other than fulfilling what they might have perceived to be a professional obligation. These participants tended to be on the technology or instrumentation side of research. The remaining participants exhibited various motivations, including a personal interest in the topic of my research and in helping me gather the information I needed to proceed. It was this last group that was particularly helpful to me as I began to formulate this speech code of science and I am immensely grateful to them for their openness and generosity.

PUTTING IT ALL TOGETHER

In closing, I include an excerpt as representative of the perspectives indicated in the data as a whole. In the following excerpt, the researcher responds to my question about how to improve consideration of ethics in nanotechnology research, a lack of which the researcher had earlier described as troubling:

Excerpt 39

1 *(0073)* one [solution] is to realize, to have people realize that there is a lot of
2 technology out there and, you know, technology's a great thing and brings
3 important new information to people that can help them long-term. But,
4 we have to make sure that we give that knowledge in an informed way,
5 so, education's always a part of it and also having the right legal aspects
6 covered so that people are protected.

This researcher was the most open to consideration of broad SEI related to emerging science and technologies of all the researchers I interviewed for this project. During the course of our interview, this respondent had raised many issues of ethics and social justice that were posed by biotechnology and nanotechnology and expressed concern that "we often tend to think more advanced technology and not think about the ramifications of it." This heightened awareness, notwithstanding, the way in which this researcher characterized these technologies, the role of the scientist, and the role of communication remained consistent with what I found across the data, suggesting a strong set of shared norms pertaining to communicative conduct among scientists.

Regarding the nature of scientific and technological advancement, the interview data indicated that the scientists and engineers I spoke to understand this nature in both deterministic (e.g., one cannot stop scientific and technological advancement) and beneficent (e.g., scientific and technological advancement improve the world) terms that I am suggesting may preclude full consideration of broad normative questions of social and ethical impact. Although not explicit in the aforementioned excerpt, the former characterization is implicit in this researcher's understanding of how to confront social and ethical challenges posed by emerging technologies, namely by informing and protecting the public (lines 4–6). No suggestion that as a society we have a choice to determine which technologies we pursue was indicated. The latter characterization is evident in lines 2–3 in the aforementioned quoted excerpt where the researcher characterizes technology as "a great thing" that "brings important new information to people that can help them long-term."

Regarding the appropriate relationship between scientist and society, those interviewed characterized this relationship in terms of "public education"; that is, educating the public on the benefits of emerging science and technology. Public concerns were characterized as ridiculous and inaccurate, based on science fiction portrayals or other misinformation or simply lacking information altogether. Additionally, the need to educate the public was characterized as a secondary need motivated by the primary need of ensuring that scientific and technological advancement was not impeded. Although these elements are absent in the excerpt I present here, the researcher references the relationship between scientist and the public as one in which scientists impart knowledge to the public (lines 4–5) that is consistent with the "public education" concept referenced throughout the data.

Finally, in the excerpt I presented earlier, the researcher discusses communication with the public in monologic terms using words such as "to have people realize" and to "give that knowledge" that implies a one-way communication model (lines 1 and 4) consistent with Borchelt and Hudson's 2008 description of standard models of communicating about science.

NOTE

1. These included one instance of the term "call" and two instances of forms of the expression "use words" ("use the word").

CHAPTER 5

Evidence for Multiple Speech Codes

Abstract Speech codes theory postulates that multiple speech codes exist in any given speech community. In this chapter, materials and data are presented that address the presence and content of additional speech codes: what speech code (or codes) are evident in the communal conversation about NSE among these scientists and engineers? Evidence is considered for additional, competing and oppositional, speech codes at work in the nano community.

Keywords Ethnography of communication · Nanotechnology · Social and ethical issues · Speech codes theory

OVERVIEW

In the previous chapters, I presented materials and data that addressed my research questions about how scientists and engineers talk about (1) their research in nanotechnology and (2) social and ethical implications of nanotechnologies and what code (or codes) were evident in the discourse as well as identified one speech code that was evident in the speech community I studied. In this chapter, I present materials and data that address the presence and content of additional speech codes indicated in my third research question:

> *RQ3:* What speech code (or codes) are evident in the communal conversation about NSE among these scientists and engineers?

Bakhtin (1982) described language as existing in a complex universe of voices, past and present, dissonant and consonant, audible and silent. According to Bakhtin, no discourse represents only itself, for it reflects all the many voices that came before it and coexist with it. Thus, if the dominant code constructed and used within the community of nanotechnology is based on assumptions about the nature of a scientist, a scientist's role in society, and the proper use of rhetoric as serving to advance the goals of science and technology (e.g., unimpeded progress, adequate funding, and public support), what additional codes are present within this community?

Proposition 2 of speech codes theory states that "In any given speech community, multiple speech codes are deployed" (p. 11). It follows from Proposition 2 of speech codes theory that there would be a minimum of two speech codes used in the scientific community in which I completed this study. Additionally, as Philipsen and Coutu (2005) noted, one of the motivations of the ethnography of speaking is "a commitment to notice and give expression to means of speaking that might previously have been rendered relatively invisible, inaudible, undervalued, or systematically ignored or suppressed" (p. 370).

I mentioned in an earlier chapter that I often felt as if the scientists and engineers I interviewed were responding to questions I had not asked, to a discourse of which I was not fully aware. As my time in the field increased, however, I began to notice evidence of this discourse and later I suggest some of the elements of this code (or codes). I am stopping short of naming another code (or codes) because I believe that additional, purposeful analyses are indicated to justify doing so. However, consistent with Proposition 2 of speech codes theory (and supported by Huspek, 1993 and Bakhtin, 1981), I am confident in asserting that multiple codes exist alongside the dominant speech code of science I presented in the previous chapters.

In this chapter, I present evidence of multiple codes in the speech community of nanoscience in which I conducted this study by (1) considering additional materials from my corpus of materials collected over the course of this 4-year investigation and (2) re-presenting materials from my previous chapters. I begin with a review of the concept of a cultural code (Philipsen, Coutu, and Covarrubias, 2005).

Cultural Codes Revisited

The concept of a cultural code in speech codes theory refers to "symbols, meanings, premises, and rules about many aspects of life" (Philipsen et al., 2005, p. 58). Philipsen, Coutu, and Covarrubias write "when we speak of a culture we speak not primarily of a time or place, but of a code that was constructed, and is used in some time or place" (2005, p. 58). This is the concept of culture that underlies both cultural communication theory and speech codes theory. Where a cultural code refers to symbols, etc. about any given subject, a speech code refers specifically to symbols, etc. about other communicative conduct. Cultural codes, and speech codes, are constructed, used, and refined through an ongoing process Philipsen refers to as a communal conversation. Societies form, it can be argued, through such communal conversations in which the participants decide collectively the form and meanings of their shared cultural codes. Consistent with Proposition 2 of speech codes theory as well as Huspek's theory of oppositional codes (1993), and Bakhtin's dialogic model (1981), these communal conversations will include other or oppositional cultural codes.

In the previous chapters, I included examples in which the respondents described various difficulties interacting with others or when these difficulties were self-evident. These included the respondent who talked about using the term "nano" strategically to get funding from sources that do not value basic research, the respondents who talked about the need for public education about science, the respondent who talked about the difficulties some scientists have when interacting with ethicists, and the different reactions to the public forum on nano at Pacific Science Center described in the previous chapter. All of these references suggest a variety of cultural codes at play in the same communal conversation about nanotechnology. Users of these various codes included funders, science educators, members of the general public, ethicists, etc. Evidence of alternate codes was apparent in interactions, or reported interactions, between scientists and nonscientists.

Throughout my fieldwork, I encountered examples of tension between what I am suggesting are competing codes of communicative conduct. That is, I observed instances where another code butted up against the science code I have presented here, resulting in observable tension between users of the two codes. In what follows I present two extended examples of this tension.

Example 1: Conversation with Graduate Students in Nano

Early on in my fieldwork I sat down in the student union building for coffee with two of my fellow graduate student colleagues in a nanotech course I was taking. I will call them Andrew and Theresa. Because they were further along in their nanotech research than many of the other students in the class, I was curious to hear from them what their views were on social and ethical implications of nanotechnologies. They had also expressed curiosity in my project. Their responses to my questions about SEI related to nanotechnologies were consistent with what I would find in my later fieldwork and interviews: statements that nanotech was no riskier or more ethically troubling than any other area of science. In fact, Andrew made the statement that it was safer working with nanotech particles with unknown properties than it was "crossing the street" (0080), a similar analogy that came up in my later interviews with the scientists. For example, a respondent said that working with carbon nanotubes was safer than "being in the kitchen, or crossing the road" (0076).

Similarly, other respondents said that there were many things to be worried about related to human and environmental risk but nanotech was not one of them. These statements were often accompanied with observable annoyance on the part of the speaker, revealed in such statements as "Why just nano? There's a lot more dangerous–in fact if you ask me, there's nothing to the alleged quote, unquote, dangers of nano" (0060). I have already presented the element of the science code which says that "a scientist should not emphasize 'negative' aspects of his/her work when publishing or presenting" and suggested that this supports a premise that underlies the code that "public concern about science is often based on misunderstanding science and this misunderstanding leads to impeding scientific and technological progress." I am suggesting that the aforementioned examples are evidence of that rule and premise in discursive action. I also raise it here to show how it subtly butts up against another code, or the perception of another code. I illustrate what I mean by returning to my conversation with Andrew and Theresa.

When Andrew told me that working with nanoparticles with unknown properties was safer than crossing the street, I detected what I described in my fieldnotes as "defensiveness" from him and Theresa. I felt at the time that they were trying to convince me that nanotechnology research was a worthwhile pursuit as they told me about some of the cancer research that was being done with quantum dots and magnetic resonance imaging for early detection. This caused me to interject and say something along the lines of "Hey, I agree with you! I support science. I'm all for finding a cure for cancer–who wouldn't be?" When I said this, Theresa looked me straight in the eye and said "I am *very* glad to hear you say that" (0080). This exchange indicated to me that these two nanotech doctoral students had assumed I was antiscience based on my interest in ethics and social issues. Latour (1999) reports being surprised by such an assumption expressed by a scientist colleague, that is, the expressed assumption that Latour did not support science because he was involved in studying science (and scientists) as social phenomena. In both mine and Latour's instances, when the science code came in contact with another code (albeit a code that valued science), observable friction occurred as the users of the science code assumed that the other code users did not share their value system. In the absence of prior interactions that would have led the science code user to assume otherwise, one reason for the assumptions made is that both Latour and I engaged in discursive conduct that violated a cultural norm of the science speech code. Latour by suggesting that scientists are social beings who live in a social world (a definition that violates the code's definition of a scientist as someone who is motivated by the desire to benefit society and discover knowledge) and I by suggesting that consideration of broad SEI should be the concern of scientists which violates one of the rules I articulated from my data of this speech community that says "A scientist should not talk about broad SEI related to science."

It also bears noting here that the assumption among scientists that outsiders would be hostile to science is grounded in a history of conflict between the sciences and the humanities referred to as "the two cultures" as early as 1959 by British scientist and novelist C. P. Snow (Snow, 1998). In the mid-1990s a series of books, conferences, and special issues of journals fueled the conflict, but by the early 2000s the dust of these "science wars" had settled. I assume that many of the scientists I encountered during the course of my study might have been familiar with this conflict to varying extents; however no one I spoke to or encountered in the nano-community referenced the two-culture conflict between the sciences and the humanities. Indeed, the scientists I encountered did not report interacting with nonscientists.

Example 2: Graduate Students in NT Respond To Ethical Discussions

A second example of an instance when I observed the tension that occurred when another code came into contact with the code of science occurred in a graduate course in nanotechnology that I visited during a session on "Ethical Conduct of NT Research" which included a discussion of social and ethical implications of nanotechnology. Although not enrolled in the course, I attended this session as an observer at the invitation of the professor.

The first half of the seminar was devoted to consideration of hypothetical case studies illustrating what was labeled "general scientific conduct" ethical issues. The four cases included conflict of interest, fabrication, suspect data, and industry sponsorship. The students' responses to the cases were indicative of the overall culture of science to which I referred in the previous chapters. For example, when discussing whether it was ethical to accept industry sponsorship of one's research, the class agreed that, for the sake of promoting one's research (and, ostensibly, scientific progress), industry sponsorship was acceptable. This was consistent with the premise I later articulated that says that "strategic conduct that promotes scientific research is appropriate scientific conduct."

Overall, this section of the class proceeded quickly and smoothly, with ready answers from a number of students that seemed to indicate a high level of comfort and familiarity with the subject matter. It also supported my later findings that I presented in Chapter 4 that users of the science speech code talked about ethics in terms of internal ethical issues such as fraudulent data, etc.

The next section of the aforementioned seminar, however, entitled "Will NT 'Change the World'?" was a different matter. Again, hypothetical case studies were presented, but this time they depicted external ethical issues, the sorts of SEI related to nanotech that I had understood as SEI rather than internal ethical issues. For example, in a scenario entitled "A Taxi Driver's Dilemma," the students were presented with a hypothetical case study based on an existing nano product in which they encounter a taxi driver who, upon learning of their expertise in nanotechnology, tells them that his grandson, an aspiring gymnast in rural China, is being pressured to use a nano product that had been recently patented for dental work to strengthen his young bones. The question posed to the students is how would you advise this taxi driver? The class response to this question was complete silence. After a few uncomfortable minutes, someone suggested that "appropriate regulation is already in place" to prevent this product from being misused. Another student suggested that misuse of the product should be treated as steroids currently are, adding that "it wouldn't have been approved by the FDA" or been "released into industry" if it were not safe. These responses, as I would later discover, were consistent with what I found in my fieldwork and interviews regarding the proper role of the scientist in society. Specifically, according to this speech

EVIDENCE FOR MULTIPLE SPEECH CODES   121

code, as I presented earlier, regulation and application is not the appropriate domain (or ethical responsibility) of the scientist but of regulatory agencies and industry.

Additionally, when provided two additional examples related to use of a blood peptide, the coercion of troops to receive the peptide by the US military as a preventative measure and the misuse of the peptide by terrorists to inflict biological damage on a population, the students laughed at the examples. One student said derisively that the examples were "extreme." The professor responded that they were "required by NSF" to consider issues of ethics related to nanotechnology and although the laughter stopped, no one contributed to a discussion about the examples. The final case study in this section was that of "nanosocks," an existing product that is fabricated with nanoparticles of silver that have antimicrobial properties. The class laughed heartily at this example and several students made mocking comments regarding the idea that socks could pose a health risk. The professor responded that the same silver particles are currently being used in toothpaste and soap and that scientists do not know what happens when these particles go into our water supply and are ingested in our bodies. Again, the laughter ceased but no one responded with a comment or question.

The students' responses I have described earlier stood in marked contrast to the lively discussion that had taken place during the first half of the lecture when the internal ethical issues were presented and I interpreted the difference in the science students' response to the two sections as a difference based on a speech code. That is, the internal ethical issues were considered appropriate for them to discuss as emerging scientists; however, the external ethical issues were not considered appropriate topics for discussion and were thus met with silence and derision. Even when the professor made the mildly threatening statement that these issues were "required by NSF," the students did not respond. I suggest that this was because the topic of external ethics violated the following two rules embedded in the science speech code I presented in the earlier chapters:

"A scientist should not talk about broad social and ethical issues related to science," and "A scientist should not emphasize 'negative' aspects of his/her work when publishing or presenting."

The introduction of external SEI for discussion represented a speech code that was treated by the students as being in opposition to the one that they used. The tension that I have described earlier did not lessen until the class was over and I recall feeling discouraged by their seeming unwillingness to seriously consider macro-SEI related to nanotechnology. I will return to a discussion of this class later in this chapter but first I turn to a discussion of evidence for competing speech codes from the materials and data I presented in the earlier chapters.

Evidence of Competing Speech Codes from Earlier Materials

As I mentioned earlier, when I first began my fieldwork I attended a conference on nano-ethics attended by many prominent scholars and figures in the field of studies of nanoscience and emerging technologies. At this conference, French philosopher Jean-Pierre Dupuy called for a counter discourse to the standard discourse that addressed ethics of technologies after their development rather than before (Dupuy, 2005). Dupuy's call for a counter discourse, or speech code, one might say, has since been echoed by calls among the UK scholars in particular for "upstream" consideration of nanotechnologies rather than "downstream" (e.g., Macnaghten, Kearnes, and Wynne, 2005), with "downstream" referring to standard practices, and ways of speaking about, nanotechnologies and "upstream" referring to the counter discourse Dupuy called for. Later, I reported on an interaction that I observed at the same conference when Mihail Roco, the senior advisor for nanotechnology at the National Science Foundation and one of the most well-known and influential enthusiasts of nanotechnology, admonished a philosophy graduate student for not presenting a "scientific" paper. I noted that this outburst suggested that a code violation had occurred. However, as it turned out, Roco himself, it seems, had violated a code shared among the group of science studies scholars that dominated this particular conference. This violation was evident in the immediate response that Roco's criticism evoked among the audience. For example, Dupuy, the conference keynote speaker, responded with support of the student's philosophical premise as worthwhile. Additionally, another graduate student responded by loudly accusing Roco of co-authoring articles with a well-known transhumanist,[1] a charge that implied that Roco had his own agenda in mind when objecting to the student's questioning of nanotechnology as unscientific. Roco did not respond to either comment and the session was subsequently ended by the moderator. What Roco had derided as unscientific, the student's thesis that looking to writers such as George Orwell might be instructive in considering the ethics of present technologies, was defended as an appropriate topic for scientific discussion by members of the audience. This incident illustrated both the presence of multiple codes in the overall discussion about nanoscience and also the difficulty that some individuals, namely, Roco, had when presented with an oppositional code.

Earlier I also described how my questions about SEI related to nanotechnologies were met with silence from classmates in a nanotechnology class and, in one case, someone abruptly ending our conversation after I asked about potential applications of the research he had presented to the class. Both examples helped me understand the norms of the science speech code by creating tension with another speech code, the one I was using, even without my intentional use of it. I make this point because most of the evidence I present here for the presence of additional, opposing, or competing codes is based on the nuances of these codes within the science speech code users' talk rather than my observations of another code being used strategically.

For example, in Excerpt 1 of Chapter 2, I reported on a conversation I had with a scientist in which he said that scientists would not "have big concerns" about the social and ethical implications of nanotechnology. As I reported, I concluded that he viewed such concerns as unscientific. I am suggesting that this stance represents a discourse that is in clear opposition to another discourse that says that consideration of social and ethical implications of nanotechnologies *is* scientifically valid. Because of my affiliation with the NSF grant as an SEI researcher, I represented this discourse and so, I would argue, received the reactions I did from those I spoke to who opposed this competing way of speaking even as I was largely unaware of the particulars of either speech code.

The last example I will present from Chapter 2 is the interaction I observed between a chemist presenting his nanoscale work and a questioner who asked about application. The presenter replied "We're just chemists" (0088) and I surmised, based on the additional analyses that I presented in Chapter 2, that discussing potential applications ran counter to a scientist's identity as a basic researcher. Yet, somewhere in the question asked by the audience member, and her subsequent frustrated shrug, was the assumption that scientists *should* be able to discuss potential application of their work. Evidence for a competing speech code that carried this assumption was also found in the materials I presented in Chapter 3.

In Chapter 3 I presented Excerpt 7 in which the respondent referred to public skepticism toward scientific discoveries. I articulated a premise from this that said that "Basic science has not been valued by society as much as applied science" and another premise, "Basic science should be recognized as valuable by society."

Evidence of an oppositional code was apparent in the respondents' pejorative references to nanotechnology and interdisciplinary research, with one respondent referring disgustedly to both terms as "buzzword [s]" (0076). Thus, the oppositional code, the language of which was resentfully appropriated by this scientist and others I talked to, was that of the federal funding agencies. The reason I refer to it as oppositional rather than simply alternate is that the respondents' references to it were largely pejorative as I described in the previous chapter.

Evidence for an oppositional or competing code used by federal funding agencies can also be found in Chapter 4 where I illustrated how respondents talked about strategic conduct. In a series of excerpts with one respondent (Excerpts 4–7), I reported how the respondent talked about the need to change the way one communicates about his research in order to be responsive to funding priorities, including labeling one's research "nanotechnology." These comments clearly indicate a perceived oppositional discourse, that of funding agencies who prioritize funding to applied research over basic research. Additionally, I presented data that illustrated science code users talked about strategic conduct outside the scientific community as a one-way model of communication for explanation of, education about, and promotion of science. I contrasted this view of communication with the public with Borchelt and Hudson's 2008 call for a two-way model of public engagement with science, which represents a competing code regarding strategic conduct by scientists. I also commented on the various strategic ends those I interviewed seemed to have in participating in the interview with me, including using the interview to make an appeal for funding and using the interview to convince me that there were no SEI associated with nanotechnology. Both of these strategic aims indicate, albeit subtly, the presence of an oppositional code. Making an appeal for funding indicates opposition to a code that does not value basic research as deserving funding over applied research. Similarly, denying the existence of SEI associated with nanotechnology indicates opposition to a code that posits that these issues do exist.

It was clear to me in ways I have described earlier that those I encountered during the course of the project spoke in response to a larger discourse. As a communication researcher, I represented an oppositional code of communicative conduct without even being aware of it for the most part. The social studies of science and technology that I have referenced throughout this study also represent an oppositional code. For example, literature that urges scientists to actively consider broad, normative social and ethical considerations of their work, in large part by integrating social science and humanities research into the research and development of emerging technologies (e.g., Ebbesen, 2008; Guston and Sarewitz, 2002), indicates another way of talking about scientific identity and appropriate social relations than that illustrated by the speech of those I interviewed for this study. Additionally, literature that recasts public education efforts as public engagement, moving from a one-way information flow that emphasizes public listening to an approach that emphasizes listening on the part of the scientists (e.g., Borchelt and Hudson, 2008), is another approach to the role of strategic conduct that I have presented here.

Competing Codes, Competing Ideologies

I am not suggesting that the tension between competing ideologies I have described earlier is necessarily between only two discrete codes of communicative conduct but am suggesting that the tension I identified in this study is likely between and among multiple codes that cluster around common sets of values. Thus, one code of science represents a dominant code in use within the community of science that shares values of objectivity, rationality, and progress. Other codes represent (or suggest) other values, such as concern with the environment and social justice. I am not suggesting that this is the only dominant code within the scientific community, only that it is a dominant code that I observed during my 4 years of fieldwork and that appears to have ample evidence in a variety of private and public discursive experiences.

I am also suggesting that other codes are used not only by members of additional speech communities but by some members within the community of science, such as the senior scientist who told me off record that he was personally concerned with SEI, but he would never be able to publish on that subject in the scientific community. This statement indicated to me that the force of the code of science was such that even if a user of the code wanted to talk about social and ethical implications of science, he would not do so because of fear of being reproached by the greater community (by not being published, or by damaging his scientific reputation, presumably). Thus, while there is an awareness of other codes by this community, the overall reaction I observed was one of resistance to these codes and implying that such codes are a threat to the integrity of the community.

In "Our posthuman future: Consequences of the biotechnology revolution" (2002), American philosopher Francis Fukuyama describes the biotechnology debate as "polarized between two camps":

> The first is libertarian, and argues that society should not and cannot put constraints on the development of new technology. This camp includes researchers and scientists who want to push back the frontiers of science, the biotech industry... and... a large group that is ideologically committed to some combination of free markets, deregulation, and minimal government interference in technology. (182–183)

Indeed, Fukumaya's characterization of this "camp" is consistent with the speech code of science I have described in the present study. He describes "[t]he other camp" as "a heterogeneous group with moral concerns about biotechnology, consisting of those who have religious convictions, environmentalists..., opponents of new technology, and people on the Left who are worried about the possible return of eugenics" (p. 183). This motley assortment of individuals and groups, who would otherwise not be placed in the same category, represents the diversity of alternate, often competing, speech codes to which I refer in this study. I am suggesting that it is not possible, at this point, to identify one way of speaking about the social and ethical implications of emerging science and technologies that captures the multiple perspectives represented by a group "which ranges from activists like Jeremy Rifkin to the Catholic Church" (Fukuyama, 2002, p. 183).

Fukuyama argues that the approaches to technology development represented by the two camps he describes are "misguided and unrealistic" and he proposes international regulation of biotechnology as an alternative to the two camps (p. 183), thus adding his own "oar" to the conversation about emerging technologies (Burke, 1973). My point in citing Fukuyama here is first to show how his characterization of one "camp" (or speech code) of the biotechnology debate is consistent with the code of science I have articulated in this study and, second, to show how alternative codes exist in this conversation and indicate the complexity of articulating those codes. I would argue that multiple codes exist and were they to be identified discretely, the discourse of the given individuals and groups who employ them would need to be studied purposively and extensively as I have done here with this group of UW nanoscientists.

Although speech codes theory does not suggest that human being are "cultural automatons" (Philipsen, Coutu, and Covarrubias, 2005), but, rather, supports the notion that resisting one's own cultural code is possible, doing so is not an easy or quick process. An overly simplified reading of speech codes theory might leave a reader unjustifiably optimistic about what learning the code allows users to do. Resisting one's own cultural codes may be possible, but that does not mean it is an easy matter to do so. Deeply held values about what it means to be a person, etc. are bound up in a given speech code. Rejecting those values does not happen easily and I turn to Huspek (1993) to illustrate this point:

> For the individual speaker, rival structures can thus be said to generate colliding world views, competing ideologies, that are vying for the speakers' allegiance. While this may offer the speaker something of a choice, it may also place him or her under serious strain. For selection of any one structure and its respective colliding world view (ideology)—a prerequisite if one's words and meanings are to find social validation—necessitates selection against a rival structure. (p. 16)

I understand Huspek to be saying that choosing one "structure," or speech code, over another is not just a matter of choosing different words, but of choosing different ideologies and world views. Huspek goes so far as to characterize these different world views as "colliding" and "competing," indicating the "serious strain" that choosing between these codes causes the speaker.

The stated discomfort for the term nanotechnology by many of the scientists I interviewed serves as one example of the difficulty in changing codes. As I reported in an earlier chapter, one of the scientists I interviewed told me that many scientists have "an instinctive discomfort" with the word nanotechnology because they "[did] not consider themselves technologists but scientists." Thus the term, because of its inclusion of the word "technology," contradicts an important aspect of a scientist's identity, that scientists discover knowledge, they do not make things. This identity conflict notwithstanding, scientists working with nanoscale materials or processes have adopted the term to describe their work for various strategic reasons that I described in the previous chapter, including to attract funding. Still, this case suggests that incorporating new symbols into a speech code, particularly when the given symbol challenges the values of the speech code in which it is incorporated, is not easy. The tension I encountered in this speech community around the term "nanotechnology" suggests that code adaptation is a process that will be easier for some than others, resisted and resented by some, and will take place over a period of years, influenced by a range of changing factors that include education, priorities in funding, world events, popular culture, and so on. I now turn to three examples that illustrate how code change can occur within the nanocommunity.

Talk, Time, and Code Change

Earlier in this chapter I reported on an unsuccessful attempt I observed to engage nanotechnology graduate students in serious discussion about social and ethical implications of nanotechnologies. Two years later, I visited the same session of the aforementioned course in which I presented a lecture on social and ethical implications of nanotechnologies. The format was similar to one before, with the presentation and discussion of hypothetical case studies based on existing technologies. The reactions among the students, however, were markedly different. In contrast to the reticence and dismissiveness of the previous class, this class responded readily and with enthusiasm, debating with one another about the importance of the SEI raised by the nanotech applications and how best to address them. I will suggest later some reasons for the change between the two classes, but first I describe a third interaction with the same type of students.

A year later, the professor of the aforementioned course and I jointly offered a special topics course on interdisciplinary perspectives on ethics and nanotechnology. We were worried that the course would not attract many science and engineering students and advertised it to social science and humanities departments as well. To our surprise, nine of the twelve students who enrolled were from science and engineering departments, including the nanotechnology program. Although the class discussions were characterized by frequent disagreement about such questions as the ethical responsibility of scientists and engineers to forecast future applications of their work, for example, the dismissive attitude I had observed in the initial class several years prior was gone. Most significant in illustrating this difference were comments from two nanotech graduate students in the class that "every nano student should be required to take this class."

How does one account for the differences in the student responses over the years? I would suggest that what I have personally observed over the course of a 4-year period is a culture change taking place in science that is led by young scientists who are motivated by a growing concern with pressing global issues such as climate change and the environment, social justice, world hunger, and AIDS, and want to be actively involved, as scientists, in seeking solutions to these problems. A case in point is the widely viewed documentary on climate change presented by former US vice president Al Gore, "An Inconvenient Truth" (2006) which received two Academy Awards in 2007 and has led to global warming teach-in events on college campuses around the United States.

In 2005, I talked with a doctoral student described as "a star" by her faculty who had left the nanotech program after she said she realized that her desire to integrate social concerns into her research meant that she would be "going against the tide" of science. Additionally, she told me that she had been chided by her faculty for doing "soft research" (0079), a characterization that clearly questions the scientific value of research that involves social and ethical dimensions. Her account showed clear conflict between codes: the code of the student who was motivated by explicit social concerns and the code of science I have presented in these pages that said, in part, that consideration of broad SEI should not be the concern of scientists. Although at the time her code, which identified a scientist as someone who was explicitly concerned with SEI, caused her to be "going against the tide" of the science code, perhaps that tide is changing as more incoming scientists share her concerns. Two recent reports on ethnographic projects in nanotechnology labs support the notion that with talk and time, code change can occur in the nano community.

Mody (2008) reported on "several years" of ethnographic research taking place at Rice University's Center for Biological and Environmental Nanotechnology, saying:

> ... the presence of Kelty [an anthropologist] and his graduate students in the center sparked wide-ranging *discussions* among the center's nanoscientists about social justice and intellectual property. Those in turn led to a new research direction: an attempt to devise alternative chemical processes by which nanomaterials could be concocted by their end users, rather than solely by multinational corporations. (p. 43, italics added)

Similarly, Weil (n.d.) reported on "almost a year" (p. 8) of ethnographic work that was conducted by a philosophy of science student in a nanotech facility in Illinois:

> In time, he developed rapport with users and was able to initiate *discussion* of social and ethical issues that caught the interest of the engineers and scientists. These researchers from universities in the Chicago region began to *raise questions* of their own and bring in newspaper clippings. Eventually researchers from companies joined in. The trainer sent an email to the ethics specialist stating that his *conversations* with the student had changed his outlook. (p. 9, italics added)

I italicized the terms "discussions," "discussion," "raise questions," and "conversations" in the aforementioned excerpts to emphasize the role of two-way communication in the outcome of each case. In both cases, the conversations and discussions between the ethnographers and the scientists and engineers led to favorable outcomes, and, I would argue, a culture shift. I would also emphasize that length of time in each case, several years for the project Mody reported and nearly a year for the project that Weil reported, appeared to be a factor in the successful outcomes. As I have already described, in my own work it took 4 years to begin to see marked change in the nano community where I worked.

NOTE

1. The charge was correct as Roco has co-authored articles on the benefits of nanotechnology with William S. Bainbridge, codirector of Human-Centered Computing at NSF and a leader in the Transhumanist Movement.

CHAPTER 6

Conclusions

Abstract In this final chapter, the results of the project are reviewed and research questions revisited. Theoretical implications to speech codes theory and contributions of the work to science and technology studies and communication studies are considered. The descriptive and theoretic goals of the project are reviewed and study limitations are considered.

Keywords Ethnography of communication · Nanotechnology · Social and ethical issues · Speech codes theory

RESULTS SUMMARY

The research questions that guided this study were asked in relation to the community of nanoscientists at one particular US university. In particular, I asked:

RQ 1: How do these scientists and engineers working at the nanoscale talk about their research?

RQ 2: How do these scientists and engineers working at the nanoscale talk about social and ethical implications of nanoscale science and engineering (NSE)?

© The Author(s) 2017
D.R. Bassett, *Nanotechnology and Scientific Communication*,
DOI 10.1057/978-1-349-95201-4_6

RQ 3: What speech code (or codes) are evident in the communal conversation about NSE among these scientists and engineers?

In Chapters 2–4, I presented elements of a code of speaking about science that I constructed on the basis of my analysis and interpretation of the interview materials I collected, and that were supported further by observations and fieldwork research that I conducted. The code elements I discovered represent a dominant speech code that was used by the participants in this community, including graduate students, scientists, and engineers. This code of speaking about science contains beliefs among its users about identity (e.g., what type of person is a scientist?), society (e.g., what is a scientist's proper relationship to society?), and the role of strategic conduct in science.

IMPLICATIONS FOR EXTANT LITERATURE

In the present study I have applied speech codes theory to a speech community previously unexamined with the theory, that of science and engineering. Additionally, I have indicated that in certain speech communities that hold considerable historical power, such as science, multiple codes are not readily observable. What implication does this latter point have for considering the views of the theorists I invoked in Chapter 1, particularly Bourdieu (1991)? The ways in which the respondents talked about the strategic use of terms like "nanotechnology" or "interdisciplinary research" evoke Bourdieu's ideas about how language is produced and reproduced in order to exert power in a given society. For example, respondents reported that they used these terms because they were what the US Congress and the public wanted to hear. Thus, in Bourdieu's terms, they used these linguistic resources strategically for their purchasing power to obtain the capital they wanted, namely, funding.

The findings presented here also contribute to the extant literature in science and technology studies by suggesting additional ways in which the discourse of those working in nanoscience can be investigated beyond a straightforward analysis of interview responses. While I acknowledge the value of such approaches, I am suggesting that a cultural codes approach can provide an additional layer of analysis that helps reveal further insight into the discursive practices of some scientists and engineers working in nanotechnology.

This is illustrated by the published criticisms of Drexler, particularly in an exchange of letters between Drexler and the late Richard Smalley, a Rice University professor of Chemistry, Physics and Astronomy and a 1996 Nobel Prize in Chemistry recipient for his role in developing the buckyball, a nano-sized carbon molecule, that appeared in *Chemical & Engineering News* in 2003. In the letters, Smalley mocks Drexler's notion of molecular assemblers as unscientific, saying "[y]ou are still in a pretend world where atoms go where you want because your computer program directs them to go there" (Baum, 2003, p. 42). The published exchange ended acrimoniously with Smalley charging Drexler that "[y]ou and people around you have scared our children" (Baum, 2003, p. 42).

The present study contributes to the extant literature on rhetoric of science in at least one significant way. The various ways in which the scientists and engineers I spoke to and encountered used the term "nanotechology" to find common ground with one another, as well as with funders, evoked Ceccarelli's concept of polysemy (2001) that I reviewed in Chapter 2. Ceccarelli defined polysemy, in part, as multiple meanings for the same term and suggested that exploitation of polysemy could help begin a conversation about a certain concept disagreed upon by diverse groups. In addition to the term "nanotechnology," terms such as "interdisciplinary research and collaboration" and "public education and engagement" clearly are polysemous terms as I indicated in Chapter 4. What the present research, and cultural codes research in general, can offer rhetoric of science is a more in-depth and nuanced view of what these terms mean to those who use them and what their rhetorical aims are by using them in the ways in which they do with various interlocutors. For example, in the previous chapters I have clearly shown how diverse the meanings of just the term nanotechnology are. Until these widely divergent meanings are acknowledged, clarified, and agreed upon, I do not see how a productive conversation can take place between groups who understand the term differently. The contentious debate between Smalley and Drexler is a case in point. The two talk about nanotechnology in vastly different ways, Smalley to refer to carbon nanostructures made possible through chemistry and Drexler to refer to molecular manufacturing made possible through engineering. The two sides represented by each scientist are at both a discursive and ideological impasse as long as those discussing "nanotechnology" are talking about two (or more) widely different processes, techniques, and visions of what is feasible. Speech codes theory allows for (a) identification of specific codes and (b) appropriation of those codes, that then (c) has the potential to move the conversation forward.

Ceccarelli has shown that strategically ambiguous terms (polysemous terms) can work well in an interdisciplinary environment because each participant can approach the research with reasons that fit their own interests. My materials do not provide any refutation of that. My findings suggest that scientists accept, albeit begrudgingly, that nanotechnology means different things to different people (e.g., funders, engineers); however, in this case polysemy seems to impede the ability to communicate effectively between different members of the nano-community. The audience member who did not receive an answer to her question about application at the nanotechnology conference I reported on in an earlier chapter, the lack of truly interdisciplinary research among those in "nanotechnology," and the fruitless debate between Smalley and Drexler are all examples of how the multiple meanings different users attach to the word "nanotechnology" has impeded the ability to communicate effectively in this community.

Additionally, the present research complements Berube's 2005 analysis of nano rhetoric from a variety of stakeholders in the nano community by providing additional insight into the perspectives of one group of stakeholders, scientists, and engineers. By considering the cultural codes that underlie the language used to talk about nanotechnology, the present study provides a fuller understanding of this language (or rhetoric) and of the interlocutors who use it. To understand what one group of university scientists working in one of the top university nanotechnology research centers in the United States mean by "nanotechnology," "social and ethical issues," "interdisciplinary collaboration," and "communication" allows for more effective communication with this group of stakeholders.

For example, understanding that scientists define nanotechnology as NSE rather than as Drexler's original definition of molecular manufacturing is essential in understanding why they are not concerned about the dangers posed by molecular manufacturing. Although language associated with Drexler's vision of molecular manufacturing was used to promote nanotechnology to the Clinton administration, and resulted in the funding of the NNI, in the end no NNI funds were allocated to research in molecular manufacturing and most in the mainstream scientific community share Smalley's view that molecular manufacturing is not scientifically feasible and only serves to frighten the public. Kurzweil (2003) has suggested that Smalley's opposition to the feasibility of molecular manufacturing and ridicule of Drexler were based on Smalley's own fears that public concerns about molecular manufacturing (e.g., Drexler's grey goo scenario) would result in loss of funding for the NNI. Smalley himself said that "speculation about the potential dangers of nanotechnology threatens public support for it" (Baum, 2003).

Similarly, understanding that scientists do not consider SEI in general as relevant to their role as scientists is essential in understanding why questions and comments about such issues are not deemed appropriate. Finally, understanding that interdisciplinary collaboration refers to interactions between scientists and communication refers to selling one's work and educating the public is essential in identifying culturally appropriate ways to frame the interactions between scientists and social scientists and having a dialogue with the public. That is, understanding this speech code of science makes it clear that framing research as "interdisciplinary" is not going to achieve the desired result of encouraging interactions between scientists and social scientists and humanists.

"Communication" in this speech community does not imply dialogue, an important consideration when developing models to facilitate back and forth dialogue between scientists and the public. The ability to communicate effectively (to understand the communicative conduct of others and have one's own communicative conduct understood in the way it is intended) is one of the practical contributions of speech codes theory. Such an understanding is useful in practical and scholarly endeavors, including creating effective public engagement efforts, developing appropriate ethics curriculum for science and engineering students, and planning strategies promoting interdisciplinary collaboration between the physical sciences and the social sciences in the manner in which federal funding agencies such as National Science Foundation and National Institutes of Health describe.

IMPLICATIONS FOR PROPOSITION 6 OF SPEECH CODES THEORY

Speech codes theory answers the question, what, according to a particular speech code, is a human being? In this study I have addressed what respondents had to say about what kind of human being is a scientist. Speech codes theory also answers the question of how should human beings join other human beings in groups? In Chapter 3, I provided evidence from the speech of my respondents suggesting that certain types of interactions, such as interdisciplinary research collaboration, were appropriate between scientists only. Speech codes theory also answers the question of how to use communication strategically in order to be effective. In Chapter 4 in particular, I showed how respondents talked about the use of communication to further their goals as scientists (e.g., to secure funding).

Overall, what I found was that scientists talked about themselves in terms of working to benefit humanity and society and that, by and large, talked about appropriate interaction and communication within this speech community as that which takes place between scientists. Additionally, communication was valued as a means of explaining terms to colleagues in related fields, selling research proposals to funders, recruiting future scientists, and educating the public about the benefits of science and technology.

I was surprised to find, in interview after interview, that scientists and engineers developing nanotechnologies seemed unwilling to discuss broad normative ethical issues related to their work. After analysis of the data in toto, I concluded that a strong cultural norm that posits science and technology as serving to benefit society simply did not support a full consideration of other outcomes. This finding is consistent with Proposition 6 of speech codes theory, that "[t]he artful use of a shared speech code is a sufficient condition for predicting, explaining, and controlling the form of discourse about the intelligibility, prudence, and morality of communicative conduct" (Philipsen, Coutu, and Covarrubias, 2005, p. 23). Thus, even if a member of the speech community I investigated was concerned about broad ethical issues associated with scientific and technological development, the force of the speech code is such that it discourages the speech code user from discussing these issues in ways outside of their code. As one respondent told me, "scientists in general are always interested in bettering society so the idea that anything they could do could harm society is not a part of—cuz they spend countless hours trying to do things that would help people, not to hurt people" (0073). Considering the strength of this norm, it is perhaps not surprising that the related understandings regarding the role of the researcher vis-à-vis the public and the need for strategic communication accompanied this perspective.

Central to speech codes theory is the premise that while speech codes exert cultural force, that force is not absolute. The purpose of Proposition 6 is to address the question "what force do culture in general and speech codes in particular have in social life?" (Philipsen, 1997,

p. 146). The answer to this question is that Proposition 6 "does not generate predictions that in all circumstances interlocutors will *perform* actions that are congruent with a code, but does generate predictions about how interlocutors will *talk* about such performances" (Philipsen, 1997, p. 148, italics added). Philipsen explains that the process of discursive force described in Proposition 6 is possible because speech codes are made up of elements that "derive their meaning from their place in a network of reinforcing and interanimating code elements" (1997, p. 149). An example of this network of elements from the present study would be the symbols "science" or "scientist" deriving their meaning in part by the associated premises and rules such as "A scientist is motivated by the desire to benefit society" (P1). Philipsen also says that speech codes have the force they do because they are "socially legitimized," that is, "[t]he association of code elements, which are involved in discourse, with the memory of life experiences in these socializing contexts, endows them with a sense of legitimacy and normalcy" (1997, p. 149). Thus, the schooling of scientists and subsequent limited interactions outside their departments and disciplines create the "socializing contexts" in which the elements of a speech code are learned.

Proposition 6 "not only explains the circumstances under which the deployment of speech codes has rhetorical traction, it also provides an explanation of why they have the force they do" (1997, p. 152). Thus, Proposition 6 addresses the strength of culture and why it matters in the lives of those who utilize its linguistic (and communicative) resources. If one knows the speech code (the culture) of a group, then one may not be able to predict what people *do* but they can predict how they will *talk about* what they and others do. Knowing the code allows one to evaluate, interpret, control, predict, and explain the communicative conduct of others. One of the scientists I interviewed for this project explained the communicative conduct of other scientists, for example, what I may have perceived as "dismissiveness" of social and ethical concerns, by saying that if scientists appeared dismissive it was likely because they had already thought through the issues and come to the conclusion that there was nothing about which to be concerned.

Another National Science Foundation project dealing with SEI related to nanotechnologies conducted at Illinois Institute of Technology suggests that over time a code of science can adapt to include more direct discussion of SEI. Weil (n.d.) reported that a philosophy of science student spent a year of part-time ethnographic work in a nanotechnology facility at a university having conversations with industry users, scientists, and engineers. Over time, the student reported that the scientists and engineers he conversed with became more interested in topics of social and ethical import and even began to bring in news clippings related to these issues. These activities led to the development of an interdisciplinary course for engineering, science, and science studies students. Weil concludes that the project provides a "model for generating awareness of ethical and societal concerns among engineers in a nano facility, laboratory, or workplace" (p. 8). I would suggest that one of the factors in the success of this project was the consistent and sustained interaction that occurred with the student working over the course of a year in the user facility. In the course of the 4 years I was involved with the project I report on here, I observed positive changes in the conversations that took place about the importance of social and ethical considerations of nanotechnologies within the UWCNT. For example, the completed project led to the development of an interdisciplinary course in ethical implications of nanoscience and nanotechnology that I co-taught with a faculty member in the sciences in early 2009. The course continues to be offered. That there is sufficient interest to warrant offering the course is testament to the fact that consistent and sustained interactions have been fruitful.

GOALS OF THE PROJECT

My goals for this study were to answer three research questions and included "descriptive" and "theoretic" goals (Philipsen and Coutu, 2005, p. 357). In terms of providing descriptive value, the overarching goal of this study was to examine a communal conversation taking place about the values of technology and science in general, and about nanotechnology in particular, within a particular scientific community. Specifically, I have attempted to provide depth and insight into the perspectives of one group of participants in this communal conversation, that of scientists and engineers working at the nanoscale and to hear directly from them regarding their views on nanotechnology and the associated social and ethical implications. By presenting elements of a code of speaking about scientists, society, and the role of strategic conduct, as I have done, I have tried to meet this goal. My intent in presenting this code was not to criticize science or scientists, but rather to present a way of speaking about nanotechnology that is distinctive among this speech community. I have attempted to articulate and interpret a way of speaking about social and ethical implications of nanotechnology that seem to have some observable force among some university scientists and engineers who use this speech code.

Applying speech codes theory to a speech community previously unexamined with this theory (i.e., a speech community composed of scientists) suggests the theory's robustness and transferability to multiple contexts of human life. That the theory proved explanatory in the speech community examined here indicates future studies with speech codes theory in additional speech communities. There have been many fieldwork studies of scientific work, including ethnographic studies of scientific communities of practice, several of which I have cited in previous chapters. However, to my knowledge there are no published studies of scientific communities that explicitly use the methodological perspectives of speech codes theory, such as I have used in this study. Thus, this is the first identification of a speech code used among a speech community of scientists. As such, it adds to the hundreds of published studies in ethnography of communication that span multiple languages, cultures, and continents. It can be used to aid another ethnographer of communication who seeks to undertake a similar study, adds to the cumulative "basic knowledge about how to go about learning another culture's code for communicative conduct" (Philipsen, n.d., p. 17), and suggests how ethnography of communication and speech codes theory can contribute in a substantive way to the goals of a US federally funded research project such as the NNIN.

In terms of theoretic value, my goal was to provide additional theoretical insight into speech codes theory. In this chapter I have presented some of those insights.

Study Limitations

The most significant limitation I can identify is the length of time I was able to spend in the field. At the outset of my project I attended a lecture by an ethnographer of communication who was visiting the UW campus. I recall being surprised when he said that he spent 7 years in the field before he even began to write. After 4 years of immersion in a community of nanotechnology research, I felt that I had only just begun to grasp the complexity of this community. Suddenly, 7 years did not seem as long as it once did. Still, practical issues often dictate that one spend less time conducting one's research than is ideal. Despite the relatively short length of time spent in the field, I believe I have been able to identify significant aspects of a dominant speech code used by this speech community.

The second major limitation that I identify in this study is the fact that I do not have a background in science. Having a background in one of the physical sciences would have given me an advantage as a social sciences researcher in this community, granting me some insider knowledge about what is and is not appropriate communicative conduct and so on. Lacking this background, I had to learn by trial and error and as an outsider to this community. For example, I learned that my respondents did not like the term "nanotechnology," that they did not consider it appropriate in their role as scientists to discuss social and ethics issues related to science and technology, and that they understood the term "interdisciplinary" to refer to interactions between themselves and other scientists. However, being an outsider has certain advantages to the ethnographer of communication. The main advantage is that, lacking insider information, the ethnographer is able to observe the various elements of a speech event with minimal taken-for-granted assumptions about the premises and rules in the unfamiliar speech community. Had I been more acculturated into this community prior to beginning my fieldwork, I might have overlooked significant cultural premises and rules that would have seemed too obvious to mention. For example, I might not have asked about SEI in the first place since I would have understood that such a topic was not appropriate in this speech community.

In the ethics of emerging technologies seminar I co-taught at the UW in 2009 with graduate students from the social sciences, humanities, and physical sciences, the question was raised how do we negotiate the multiple perspectives on social and ethical implications of emerging science and technologies represented by diverse publics, not only in the United States but also around the world, to ensure that advances in nanotechnology are beneficial to all? I would suggest that speech codes theory offers one avenue for beginning to articulate the multiple perspectives that surround these issues. This study is one attempt to do so.

Appendix A: Transcription Conventions

The examples below are based on the transcription conventions found in Silverman (2003), portions of which are quoted verbatim from that text:

( )	Words inside parentheses indicate the transcriber's best estimate of what was said.
(xxx)	Repeated letter "x"s parentheses indicate unintelligible speech.
(hhh)	Repeated letter "h"s in parentheses indicate laughter.
((coughs))	Words in double parentheses indicate transcriber's comments on paralinguistic cues or other comments.
[ ]	Words in brackets indicate transcriber's attempts to clarify vague references such as "it" by inserting an earlier more specific reference made by the speaker. They are also to indicate where the transcriber has reworded an utterance in order to list utterances in a consistent way, in the same tense, for example.
(5.0)	Numbers in parentheses indicate periods of silence in seconds.
becau-	A hyphen indicates an interruption made by the speaker (the example here represents a self-interrupted "because").
<u>He</u> says	Underlining indicates stress or emphasis.

© The Author(s) 2017
D.R. Bassett, *Nanotechnology and Scientific Communication*,
DOI 10.1057/978-1-349-95201-4

BIBLIOGRAPHY

Albert, E. (1972). Culture patterning of speech behavior in Burundi. In J. Gumperz & D. Hymes (Eds.), *Directions in sociolinguistics: The ethnography of communication* (pp. 72–105). New York: Holt, Rinehart and Winston.

Alivisatos, P., Roco, M., & Williams, R. (1999). Introduction to nanotechnology for non-specialists. In *Nanotechnology research directions: IWGN workshop report: Vision for nanotechnology R&D in the next decade*, National Science and Technology Council, Committee on Technology, Interagency Working Group on Nanoscience, Engineering and Technology (IWGN).

Alivisatos, P., Roco, M., & Williams, R. (2001). Introduction to nanotechnology for nonspecialists. In *Nanotechnology research directions: IWGN workshop report: Vision for nanotechnology R&D in the next decade*, National Science and Technology Council, Committee on Technology, Interagency Working Group on Nanoscience, Engineering and Technology (IWGN).

Altmann, J. (2004). Military uses of nanotechnology: Perspectives and concerns. *Security Dialogue, 35* (1), 61–79.

Asimov, I. (1950/1991). *I, robot*. New York: Bantam Dell.

Atkinson, D. (1999). *Scientific discourse in sociohistorical context: The philosophical transactions of the Royal Society of London, 1675–1975*. Mahwah, NJ: Lawrence Erlbaum Associates.

Bakhtin, M. (1981). *The dialogic imagination*. Holquist, M. (Ed.). Austin, TX: University of Texas Press.

Bakhtin, M. (1982). *The dialogic imagination*. Holquist, M. (Ed.). Austin, TX: University of Texas Press.

Bassett, D. (2010). Norio Taniguchi. In D. Guston (Ed.) *Encyclopedia of nanoscience and society*. Thousand Oaks, CA: Sage Publications.

Baum, R. (2003). Drexler and Smalley make the case for and against "molecular assemblers." *Chemical & Engineering News, 81*(48), 37–42.

Bauman, R. (1970). Aspects of 17th century Quaker rhetoric. *Quarterly Journal of Speech, 56,* 67–74.

Bennett, I., & Sarewitz, D. (2006). Too little, too late? Research policies on the societal implications of nanotechnology in the United States. *Science as Culture, 15*(4), 309–325.

Bernton, H., & Clarridge, C. (2006). Earth liberation front members plead guilty in 2001 firebombing. The Seattle Times, October 5, 2006. Retrieved March 4, 2008, from http://seattletimes.nwsource.com/html/localnews/2003289715_uwfire05m.html.

Berube, D. (2006). *Nanohype: The truth behind the nanotechnology buzz.* Amherst, NY: Prometheus Press.

Bone, J. (2000, March 15). New sciences 'threaten end of humanity.' *The London Times.* Retrieved April 29, 2008, from Lexis Nexis Academic Database.

Borchelt, R., & Hudson, K. (2008, April 21). Engaging the scientific community with the public. *Science Progress.* Retrieved June 12, 2008, from http://www.scienceprogress.org/2008/04/engaging-the-scientific-community-with-the-public/.

Bourdieu, P. (1991). *Language and symbolic power.* Cambridge, MA: Harvard University Press.

Briggs, C. (1986). *Learning how to ask: A sociolinguistic appraisal of the role of the interview in social science research.* Cambridge, UK: Cambridge University Press.

Bullis, K. (2008). Federal research funding cut: Financial support for a major international fusion project is one of many casualties. *Technology Review.* Retrieved March 6, 2008, from http://www.technologyreview.com/Biztech/20085/.

Burke, K. (1973). *The philosophy of literary form: Studies in symbolic action.* Berkeley, CA: University of California Press.

Campbell, N. (1952). *What is science?* New York: Dover Publications.

Campbell, J. (1986). Scientific revolution and the grammar of culture: The case of Darwin's origin *The Quarterly Journal of Speech, 72*(4), 351–376.

Carbaugh, D. (1984). On persons, speech, and culture: American codes of "self," "society," and "communication" on Donahue. University of Washington doctoral dissertation.

Carbaugh, D. (1988). *Talking American: Cultural discourses on Donahue.* New York: Ablex Publishing Corporation.

Carbaugh, D. (1993). "Soul" and "self": Soviet and American cultures in conversation. *The Quarterly Journal of Speech, 79*(2), 182–200.

Carbaugh, D. (2005). *Cultures in conversation.* Mahwah, NJ: Lawrence Erlbaum Associates.

Ceccarelli, L. (2001). *Shaping science with rhetoric: The cases of Dobzhansky, Schrodinger, and Wilson*. Chicago: The University of Chicago Press.

Center on Nanotechnology and Society (CNS). (2007). Retrieved January 31, 2007, from http://www.nano-and-society.org/.

Coenen, C. (2005, March). Ethical aspects of technological convergence—Anti-utopia revisited. Paper presented at University of South Carolina, Nano-Ethics conference, Columbia, SC.

Colvin, V. (2003). The potential environmental impacts of engineered nanomaterials. *Nature Biotechnology, 21*(10), 1166–1170

Coutu, L. (2000). Communication codes of rationality and spirituality in the discourse of and about Robert S. McNamara's in retrospect. *Research on Language and Social Interaction, 33*, 179–212.

Covarrubias, P. (2002). *Culture, communication, and cooperation: Interpersonal relations and pronominal address in a Mexican organization*. Lanham, MD: Rowman & Littlefield.

Crichton, M. (2002). *Prey*. New York: HarperCollins.

Deepening Ethical Engagement and Participation in Emerging Nanotechnologies (DEEPEN). (2007). Retrieved September 17, 2008, from http://www.geography.dur.ac.uk/projects/deepen/Home/tabid/1871/Default.aspx.

Dirven, R., Goossens, L., Putseys, Y., & Vorlat, E. (1982). *The scene of linguistic action and its perspectivization by speak, talk, say and tell*. Philadelphia: John Benjamins.

Drexler, K.E. (1986). *Engines of creation: The coming era of nanotechnology*. New York: Anchor Books.

Dupuy, J. (2005, March). The philosophical foundation of nanoethics: Arguments for a method. Paper presented at the conference on NanoEthics at the University of South Carolina, Columbia, SC.

Ebbesen, M. (2008). The role of the humanities and social sciences in nanotechnology research and development. *Nanoethics, 2*, 1–13.

ETC Group. (2003). News release: More evidence for moratorium on synthetic nanoparticles. Retrieved May 1, 2008, from http://www.etcgroup.org/en/materials/publications.html?id=164.

ETC Group. (2006). Nanotech product recall underscores need for nanotech moratorium: Is the magic gone? Retrieved March 6, 2008, from http://www.etcgroup.org/en/materials/publications.html?pub_id=14.

Feldman, S. (2004). The culture of objectivity: Quantification, uncertainty, and the evaluation of risk at NASA. *Human Relations, 57*(6), 691–718.

Feynman, R. (1960). There's plenty of room at the bottom. Retrieved March 31, 2008, from http://www.zyvex.com/nanotech/feynman.html.

Fleck, L. (1979). *Genesis and development of a scientific fact*. T. J. Trenn & R. K. Merton (Eds.). (F. Bradley & T. J. Trenn, Trans.). Chicago: University of Chicago Press.

Fukuyama, F. (2002). *Our posthuman future: Consequences of the biotechnology revolution*. New York: Farrar, Straus and Giroux.
Gilbert, N. (2007). Academics warn of physics funding crisis. *The Guardian*. Retrieved March 6, 2008, from http://www.guardian.co.uk/education/2007/dec/11/research.highereducation.
Gilbert, G., & Mulkay, M. (1984). *Opening Pandora's box: A sociological analysis of scientists' discourse*. Cambridge, UK: Cambridge University Press.
Gould, S. J. (2003). *The hedgehog, the fox, and the magister's pox: Mending the gap between science and the humanities*. New York: Harmony Books.
Guston, D. (2010). *Encyclopedia of nanoscience and society*. Thousand Oaks, CA: Sage Publications.
Guston, D., & Sarewitz, D. (2002). Real-time technology assessment. *Technology in Society, 24*, 93–109.
Herrick, C. (2008). The Southern African famine and genetically modified food aid: The ramifications of the United States and European Union's trade war. *Review of Radical Political Economics, 40*(1), 50–66. Retrieved May 7, 2008, from http://rrp.sagepub.com/cgi/reprint/40/1/50.pdf?ck=nck
Huspek, M. (1993). Dueling structures: The theory of resistance in discourse. *Communication Theory, 3*, 1–25.
Hymes, D. (1962). The ethnography of speaking. In T. Gladwin & W.C. Sturtevant (Eds.), *Anthropology and human behavior* (pp. 13–53). Washington, D.C.: Anthropological Society of Washington.
Hymes, D. (1972). Models of the interaction of language and social life. In J. Gumperz & D. Hymes (Eds.), *Directions in sociolinguistics: The ethnography of communication* (pp. 35–71). New York: Holt, Rinehart and Winston.
Hymes, D. (1974). The contribution of poetics to sociolinguistic research. In *Foundations in sociolinguistics: An ethnographic approach* (pp. 135–141). Philadelphia: University of Pennsylvania Press.
Jeliniski, L. (1999). Biologically related aspects of nanoparticles, nanostructured materials and nanodevices. In R.W. Siegel, E. Hu, & M.C. Roco (Eds.), *Nanostructure science and technology: A worldwide study* (pp. 113–130). National Science and Technology Council (NSTC), Committee on Technology, and The Interagency Working Group on NanoScience, Engineering and Technology (IWGN). Retrieved January 19, 2009, from http://www.wtec.org/loyola/nano/IWGN.Worldwide.Study/ch7.pdf.
Johnson, D. (2007). Ethics and technology 'in the making': An essay on the challenge of nanoethics. *Nanoethics, 1*, 21–30.
Johnstone, B. (2002). *Discourse analysis*. Malden, MA: Blackwell.
Jotterand, F. (2005, March). The politicization of science and technology: Its implications for nanotechnology. Paper presented at University of South Carolina Nano-Ethics Conference, Columbia, SC.

Joy, B. (2000). Why the future doesn't need us. *Wired*, *8*(04), 1–11. Retrieved January 14, 2009, from http://www.wired.com/wired/archive/8.04/joy.html.
Juengst, E. (1996). Self-critical federal science? The ethics experiment within the U.S. Human Genome Project. *Social Philosophy and Policy*, *13*(2), 63–95.
Kahane, D. (2003). Dispute resolution and the politics of cultural generalization. *Negotiation Journal*, *19*(1), 5–27.
Katriel, T., & Philipsen, G. (1981). "What we need is communication": "Communication" as a cultural category in some American speech. *Communication Monographs*, *48*(4), 301–317.
Kearnes, M. (n.d.). (Re)Making matter: Design and selection. Unpublished manuscript. Retrieved January 15, 2006, from http://www.geography.dur.ac.uk/projects/deepen/Outputs/tabid/1994/Default.aspx. 1–24.
Kearnes, M., Grove-White, R., Macnaghten, P., Wilsdon, J., & Wynne, B. (2006). From bio to nano: Learning lessons from the UK agricultural biotechnology controversy. *Science as Culture*, *15*(4), 291–307.
Keating, E., & Jarmon, L. (2006). What is nanotechnology: New properties of words as territories in a cross-disciplinary, cross border flow. *Practicing Anthropology*, *28*(2), 6–10.
Kinsella, W. (1999). Discourse, power, and knowledge in the management of "big science": The production of consensus in a nuclear fusion research laboratory. *Management Communication Quarterly*, *13*, 171–208.
Kuhn, T. (1996).*The structure of scientific revolutions*. Chicago, IL: University of Chicago Press.
Kurzweil, R. (2000). *The age of spiritual machines: When computers exceed human intelligence*. New York: Penguin.
Kurzweil, R. (2003). The Drexler Smalley debate on molecular assembly. Retrieved April 24, 2009, from http://www.kurzweilai.net/meme/frame.html?main=/articles/art0604.html.
Kurzweil, R. (2005). *The singularity is near: When humans transcend biology*. New York: Penguin Books.
Latour, B. (1999). *Pandora's hope: Essays on the reality of science studies*. Cambridge, MA: Harvard University Press.
Leading technologist warns about computer-takeover of planet. (2000, March 13). *Deutsche Presse-Agentur*. Retrieved April 29, 2008, from Lexis Nexis Academic Database.
Leshner, A. (2006). Science and public engagement. *The Chronicle Review*, *53*(8), B20. Retrieved April 6, 2017, from http://chronicle.com/weekly/v53/i08/08b02001.htm.
Leighter, J. (2007). Codes of commonality and cooperation: Notions of citizen personae and citizen speech codes in American public meetings. University of Washington doctoral dissertation.

Macnaghten, P., Kearnes, M., & Wynne, B. (2005). Nanotechnology, governance, and public deliberation: What role for the social sciences? *Science Communication*, 27(2), 1–24.

Madison's Nano Cafés. (2007). Retrieved January 31, 2007 from, http://www.nanocafes.org/.

Markoff, J. (2000, March 13). Technologists get a warning and a plea from one of their own. *The New York Times*, p. C1. Retrieved April 29, 2008, from Lexis Nexis Database.

Markoff, J. (2000, March 14). The risks of things to come: Breaking ranks with the high-tech optimists, Sun Microsystems scientist Bill Joy warns that 21st century technologies like robotics, nanotechnology and genetics can spawn new classes of accidents and abuses. *National Post*. Retrieved April 29, 2008, from Lexis Nexis Database.

McCann-Mortimer, P., Augoustinos, M., & LeCouteur, A. (2004). 'Race' and the human genome project: Constructions of scientific legitimacy. *Discourse & Society*, 15(4), 409–432.

McKibben, B. (2003). Enough: Staying human in an engineered age. NewYork: Owl Books.

Mody, C. (2008). The larger world of nano. *Physics Today*, 61(10), 38–44.

Nanoscale Science, Engineering and Technology Subcommittee (NSET), Committee on Technology (CT), & National Science and Technology Council (NSTC). (2004). *The National Nanotechnology Initiative strategic plan*, i–31.

National Nanotechnology Infrastructure Network (NNIN). (2008a). Retrieved May 1, 2008, from http://www.nnin.org/nnin_society_ethics.html.

National Nanotechnology Infrastructure Network (NNIN). (2008b). Societal and ethical implications of nanoscale science and engineering. Retrieved May 1, 2008, from http://www.nnin.org/doc/TrainingSlidesv6.pdf.

National Nanotechnology Infrastructure Network (NNIN). (2008c). SEI research: Current research activities, Retrieved April 29, 2008, from http://www.sei.nnin.org/sei_research.html.

National Nanotechnology Initiative (NNI). (2007). Retrieved January 31, 2007, from http://www.nano.gov/index.html.

National Nanotechnology Initiative (NNI). (2009). Retrieved February 23, 2009, from http://www.nano.gov/html/about/funding.html.

Paul, P. (2005). Exploring the social and ethical dimensions of nanofabrication at CNF. Retrieved February 11, 2008, from http://www4.cnf.cornell.edu/Ethics/reu.wmv.

Philipsen, G. (n.d.). Researching culture in contexts of social interaction: An ethnographic approach, a network of scholars, illustrative moves. Unpublished manuscript, 1–27.

Philipsen, G. (1972). Navajo world view and culture patterns of speech: A case study in ethnorhetoric. *Speech Monographs, 39*(2), 132–139.

Philipsen, G. (1975). Speaking "like a man" in Teamsterville: Culture patterns of role enactment in an urban neighborhood. *The Quarterly Journal of Speech, 61,* 13–22.

Philipsen, G. (1976). Places for speaking in Teamsterville. *The Quarterly Journal of Speech, 62*(1), 15–25.

Philipsen, G. (1986). Mayor Daley's council speech: A cultural analysis. *The Quarterly Journal of Speech, 72,* 247–260.

Philipsen, G. (1987). The prospect for cultural communication. In L. Kincaid (Ed.), *Communication theory from eastern and western perspectives* (pp. 245–254). New York: Academic Press.

Philipsen, G. (1992). *Speaking culturally: Explorations in social communication.* Albany, NY: State University of New York Press.

Philipsen, G. (1997). A theory of speech codes. In G. Philipsen & T. Albrecht (Eds.), *Developing communication theories* (pp. 119–156). Albany, NY: State University of New York Press.

Philipsen, G. (2000). Permission to speak the discourse of difference: A case study. *Research on Language and Social Interaction, 33,* 213–234.

Philipsen, G. (2002). Cultural communication. In Gudykunst, W. & Mody, B. (Eds.), *Handbook of international and intercultural communication* (pp. 51–67). Thousand Oaks, CA: Sage Publications.

Philipsen, G., & Carbaugh, D. (1986). A bibliography of fieldwork in the ethnography of communication. *Language in Society, 15,* 387–398.

Philipsen, G., & Coutu, L. (2005). The ethnography of speaking. In K. Fitch & R. Sanders (Eds.), *Handbook of language and social interaction* (pp. 355–379). Mahwah, NJ: Lawrence Erlbaum Associates.

Philipsen, G., & Leighter, J. (2007). Sam Steinberg's use of "tell" in *Corporation: After Mr. Sam*. In F. Cooren (Ed.), *Interacting and organizing: Analyses of a management meeting* (pp. 205–223). Mahwah, NJ: Lawrence Erlbaum Associates.

Philipsen, G., Coutu, L., & Covarrubias, P. (2005). Speech codes theory: Restatement, revisions, and response to criticisms. In W. B. Gudykunst (Ed.), *Theorizing about intercultural communication.* Thousand Oaks, CA: Sage Publications.

Ratner, M., & Ratner, D. (2003). *Nanotechnology: A gentle introduction to the next big idea.* Upper Saddle River, NJ: Pearson Education, Inc.

Rejeski, D. (2008). The tower of nano babel or how high-tech hucksterism can hurt nanotechnology's future. *Nanotechnology Now.* Retrieved March 6, 2008, from http://www.nanotech-now.com/columns/?article=174.

Rhodes, R. (1986). *The making of the atomic bomb.* New York: Simon & Schuster.

Schatzman, L., & Strauss, A. (1973). *Field research: Strategies for a natural sociology.* Englewood Cliffs, NJ: Prentice-Hall.

Scheufele, D. (2008). Religion and nano: What the data show. *Nanopublic.* Retrieved May 15, 2008, from http://nanopublic.blogspot.com/2008/02/religion-and-nano-what-data-show.html.

Schummer, J. (2004). Multidisciplinarity, interdisciplinarity, and patterns of research collaboration in nanoscience and nanotechnology. *Scientometrics, 59*(3), 425–465.

Schummer, J. (2005, March). Ethics in the age of fundamentalism: Vacillating between nano-salvation and nano-Armageddon. Paper presented at University of South Carolina Nano-Ethics conference, Columbia, SC.

Schummer, J. (2007). Impact of nanotechnologies on developing countries. In F. Allhoff, P. Lin, J. Moor, & J. Weckert (Eds.), *Nanoethics: The ethical and social implications of nanotechnology* (pp. 291–307). Hoboken, NJ: John Wiley & Sons, Inc.

Selzer, J. (1993). *Understanding scientific prose.* Madison, WI: University of Wisconsin Press.

Silverman, D. (2003). *Qualitative research: Theory, method and practice.* London: Sage Publications.

Small, H., Kushmerick, A., & Benson, D. (2008). Scientists' perceptions of the social and political implications of their research. *Scientometrics, 74*(2), 207–221.

Snow, C. P. (1998). *The two cultures.* Cambridge, UK: Cambridge University Press.

Sparrow, R. (2007). Revolutionary and familiar, inevitable and precarious: Rhetorical contradictions in enthusiasm for nanotechnology. *Nanoethics, 1,* 57–68.

Talbot, D. (2006). Good news: No nano news. *Technology Review.* Retrieved March 6, 2008, from http://www.technologyreview.com/blog/editors/17430/.

Taylor, C. (1996). *Defining science: A rhetoric of demarcation.* Madison, WI: University of Wisconsin Press.

The Project on Emerging Nanotechnologies (PEN). (2008). Nanotechnology consumer products inventory. Retrieved May 1, 2008, from http://www.nanotechproject.org/inventories/consumer/.

THOMAS. (2008). *21st Century Nanotechnology Research and Development Act.* Retrieved September 30, 2008, from http://thomas.loc.gov/. 1–10.

Torgersen, H., Hampel, J., von Bergmann-Winberg, M., Bridgman, E., Durant, J., Einsiedel, E., et al. (2002). Promise, problems and proxies: Twenty-five years of debate and regulation in Europe. In M. Bauer & G. Gaskell (Eds.), *Biotechnology: The making of a global controversy* (pp. 21–94). Cambridge, UK: Cambridge University Press.

Uldrich, J. (2007, August 27). Nano's big numbers. *The Motley Fool.* Retrieved September 18, 2007, from http://www.fool.com/investing/general/2007/08/27/nanos-big-numbers.aspx.

Webb, J., Schirato, T., & Danaher, G. (2002). *Understanding Bourdieu*. London: Sage Publications.

Weil, V. (n.d.). Oversight of nanotechnologies: Why this is the right time for voluntary standards of care. Unpublished manuscript.

Weiss, R. (2008a, May 21). Effects of nanotubes may lead to cancer, study says. *Washington Post*, p. A02.

Weiss, R. (2008b). Groups petition EPA to ban nanosilver in consumer goods. *Washington Post*. Retrieved July 9, 2008, from http://www.washingtonpost.com/wp-dyn/content/article/2008/05/01/AR2008050103228.html.

Worthen, B. (2008). Nanotechnology is morally unacceptable. *The Wall Street Journal*. Retrieved May 15, 2008, from http://blogs.wsj.com/biztech/2008/02/21/nanotechnology-is-morally-unacceptable/?mod=homeblog mod_businesstechnology.

Index

A
The atomic force microscope, 2

B
Bakhtin, 14, 16, 17, 18, 116, 117
Bourdieu, 14, 15, 18, 132
The buckyball, 2
Burke, Kenneth, 16

C
Communal conversations, 11, 117
Communication, 10, 78, 135
Communicative conduct, 10–15, 24–26, 34, 55, 56, 78, 81, 89–91, 97, 106, 109–112, 117, 125, 135–137, 139–140
Cultural codes, 11, 14, 16, 55, 117, 127, 132–134
Cultural communication theory, 10, 11, 117
Culture, 10, 12, 14, 16, 21, 26, 45, 53, 85, 86, 97, 117, 119, 120, 127, 128, 129, 136, 137, 139

D
Drexler, Eric, K., 2, 3, 22n1, 68, 79–82, 84, 133–135

E
Ethnography of communication, 10, 23, 24, 139
The ethnography of speaking, 16, 17, 116

F
Feynman, Richard, 1, 2
Fukuyama, 126

G
Gray Goo, 2, 3

H
The Human Genome Project, 8
Hymes, 10, 13, 16, 17, 21, 24, 27

I

Interdisciplinary, 25, 87, 88, 91, 95, 97, 103, 105, 107, 124, 128, 132–135, 138, 140

J

Joy, Bill, 2, 3, 79, 80

K

Kurzweil, 3, 5, 19, 28, 79, 80, 81, 84, 134

N

Nanoethics, 8
Nanoscale science and engineering, 1, 3, 4, 18, 23, 78, 131, 134
Nanoscience, 4, 7, 8
Nanoscience and nanotechnologies, 1, 17
Nanotechnology, 1–10, 13, 14, 19–20, 22n3, 23–35, 38–41, 43, 46–47, 53, 56, 61, 67–75, 78–85, 88–91, 98, 104, 106–108, 110–112, 116–124, 127–129, 130n1, 132–135, 138–140
National Nanotechnology Infrastructure Network, 6, 20, 139
The National Nanotechnology Initiative, 5, 6, 79, 134

O

Oppositional cultural codes, 117

P

Philipsen, Gerry, 10–14, 16–18, 24, 26–28, 45, 55, 56, 77–78, 90, 97–98, 103–104, 116–117, 127, 136–139
The Project on Emerging Technologies, 7
Psychology, 12, 27, 28
Public education about science, 117
Public engagement, 91, 92, 93, 107, 108, 124, 125, 135

R

Rhetoric, 8, 11, 12, 13, 17, 25, 27, 77–84, 86, 87, 89, 116, 133–134
Roco, 4, 26, 79, 82, 83, 122, 130n1

S

The science code, 78, 117–119, 129
Science communication, 91
SEI, 5, 6, 7, 120, 123
Social and ethical issues, 2, 4, 7
The Society for the study of Nanoscience and Emerging Technologies, 7
Sociology, 12
SPEAKING framework, 13, 14, 16, 21, 41, 81
Speech code, 10–18, 23–28, 55, 56, 61, 66, 68, 73–75, 77, 78, 82, 93, 97, 98, 103–104, 108, 111, 115–117, 119–123, 126–128, 132, 135–137, 139–140
Speech codes theory, 10–18, 26–28, 56, 116–117, 127, 132, 135–136, 139–140

Speech community, 11–13, 16–17, 23–32, 35, 42–43, 45, 54, 58–59, 61, 65–66, 68, 72, 74–75, 78, 83, 88–91, 97, 106, 111, 115–116, 119, 128, 132, 135–136, 138–140
Strategic conduct, 13, 25, 27, 77–79, 85, 88–90, 93, 97, 108, 120, 124, 125, 132, 138

T

The 21st Century Nanotechnology Research and Development Act, 2, 5

CPSIA information can be obtained
at www.ICGtesting.com
Printed in the USA
BVOW06*0505130617
486752BV00004BA/45/P